W9-BBC-713

Why Care for Nature?

The International Library of Environmental, Agricultural and Food Ethics

VOLUME 9

WHY CARE FOR NATURE?

IN SEARCH OF AN ETHICAL FRAMEWORK FOR ENVIRONMENTAL RESPONSIBILITY AND EDUCATION

by

Dirk Willem Postma

A C.I.P. Catalogue record for this book is available from the Library of Congress.

ISBN-10 1-4020-5002-X (HB)
ISBN-13 978-1-4020-5002-2 (HB)
ISBN-10 1-4020-5003-8 (e-book)
ISBN-13 978-1-4020-5003-9 (e-book)

Published by Springer,
P.O. Box 17, 3300 AA Dordrecht, The Netherlands.

www.springer.com

Printed on acid-free paper

CONTENTS

ABOUT THE AUTHOR

Dirk Willem Postma was born on 21 August 1973 in Gorredijk, the Netherlands. After completing his secondary education in 1990 he studied at Teacher Training College for Primary Education in Drachten. Having obtained his teaching qualification in 1994 he moved to Groningen to study Philosophy and History of Education at the Department of Education of the State University of Groningen, where he graduated in 1997. During his college days he participated in several environmental organisations and local pressure groups. From 1999 to the end of 2004 he worked as a PhD-student at the Institutes of Philosophy and History of Education of the Radboud University of Nijmegen, the Netherlands, and the Catholic University of Leuven, Belgium. His study deals with the ethical and politico-philosophical dimensions of environmental responsibility and education. Dirk Willem Postma was editor of *Vernieuwing* (*Reform*), a progressive magazine for education and teaching from 1998 until 2005. Since spring 2005 he works as editor for *SWP Publishers*, a Dutch publishing house of books and magazines in the field of education, youth care, social work and humanism. As such, he is editor of the educational magazine *Pedagogiek in Praktijk* (*Education in Practice*).

ACKNOWLEDGEMENTS

Having written in virtual seclusion this past year has nevertheless given me a deeper appreciation of the philosophical commonplace that authors are not the intellectual owners of the texts they write. Obviously, this book has been written by me, in the sense that the touch of my fingers on the keyboard caused the text to appear on the computer screen. But on a more intellectual level, this book ought to be read as the result of an ongoing conversation with others on issues of shared interest, or even as a testimony of sharing a common life with some of them. Therefore, I would personally like to thank all these 'others' for their part in the creation of this dissertation. However, I do feel that there are more proper ways to express *personal* gratitude than a few sentences on the first page of an academic monograph such as this. Let me suffice here by confining myself to a rather compact list of acknowledgements.

First of all I would like to thank my supervisors Paul Smeyers and Wouter van Haaften for their professional supervision, inspiring and meticulous criticism and personal involvement. Furthermore, I am grateful to my colleagues at the Department of Philosophy of Education in Leuven as well as in Nijmegen, in particular Ger Snik and Johan de Jong who ought to be mentioned for their contribution to my research plan and criticism of the initial texts. I am most grateful to my roommate and friend Femke Takes for her solidarity and our sharing of the ups and downs in the mad life of a junior researcher. I want to thank the participants of the Dutch Kohnstamm Network and the participants of the Leuven Research Community for the stimulating debates on relevant issues in our field of study. I would also like to express my thankfulness to the senior colleagues of different universities who were prepared to comment on early drafts of my papers on several occasions. Wilna Meijer, Bas Levering, Jan Bransen, Paul Standish, Richard Smith, Michael Bonnett, Stefaan Cuypers and Hub Zwart deserve special mention – as well as those junior colleagues who were never tired of discussing social, political and philosophical issues in a personal and generous way. I want to thank Scott Rollins for his careful editing of my manuscript. My fellow editors of the educational magazine *Vernieuwing (Reform)* kept me from becoming narrow-minded by widening my horizon of concern to issues and opinions beyond the immediate scope of my research. And of course, there are my friends who have endured my moods and absent-mindedness, and who have – each in their own personal way – showed me that there are worthwhile things to do and experience outside of my study. In doing so they regenerated and strengthened my determination to go on. Most of all I thank my parents, Pier and Willy Postma, for their unconditional faith and support without which I would never have been able to accomplish this. Finally, I am grateful beyond words to my partner Niek. He knows why.

Dirk Willem
Amsterdam, November 2004

CHAPTER ONE

INTRODUCTION

Don't you see that the whole aim of Newspeak is to narrow the range of thought?

George Orwell[1]

This book has been written in a time of environmental neglect. A time in which the expansive needs of multinational corporations, western consumer interests and the politically celebrated ideals of economic growth and technological progress appear to override any consideration for preserving natural beauty as well as consideration for those unable to speak and negotiate on their own behalf: third world citizens, future generations, animals, plants and landscapes. This neglect is evident in the withdrawal of national governments from the requirements of international agreements on the reduction of greenhouse gas emissions (Kyoto treaty), it is manifest in the organised inability and unwillingness to establish more equal trade relations between rich and poor countries as well as in the lack of political commitment to protect extraordinary sites of natural beauty from economic exploitation (Alaska, the Amazon rainforests, the Dutch Wadden Sea). In times like these, environmental education is a hazardous and primarily ambiguous enterprise, since it easily comes to function as a means to foist present responsibilities onto future generations. Some proponents, for instance, argue that environmental education should 'create a new generation of citizens who are greener than their parents' (Bell, 2004, p. 43). Thus, new born citizens are burdened with environmental responsibilities that we failed to live up to ourselves.

This predicament prompts careful reflection on the nature and status of environmental education: if environmental education is designed to create 'green citizens' and advance a 'sustainable development', how should we judge the prespecified aims of 'citizenship' and 'sustainable development'? And if environmental education is meant to raise aesthetic appreciation and care for the natural environment in our every day behaviour, how do we judge intrinsic value-claims and ideals of environmental responsibility? These questions will be at the heart of this PhD-thesis, which is dedicated to examining the ethical and politico-philosophical dimensions of environmental responsibility and environmental education. To be more specific, the focus of this study will be on the framework of *Education for Sustainable Development* (ESD), which dominates at present in all countries who participate in the educational and environmental organisations of the United Nations, (who recently declared the upcoming decade of 2005–2015 as the 'United Nations Decade on Education for Sustainable Development').

Some preliminary remarks should be made concerning the scope of this inquiry. To begin with, I would like to stress that my focus will be on discourses of environmental education in Western Europe and the English speaking world, because the practices of environmental education in these countries share a similar predicament, whereas consideration given to practices of environmental education in Africa, Asia or Latin-America would require knowledge of their particular school systems, curricula and educational traditions. This is not to say that my analysis of the environmental crisis and issues of sustainable development will be restricted to the western world. On the contrary, the global nature of environmental and developmental problems does not allow for such a limited view. Secondly, my primary focus will be on the practice of environmental education within formal schooling settings. This means that educational initiatives concerning sustainable community building and neighbourhood projects are not the main concern of the philosophical consideration employed here. Again, the reasons for this limitation are to be found in the nature of that which is being considered: state-initiated policies of environmental education require politico-philosophical reflection of a different kind than initiatives of social movements and practices of adult education.

If there is anything characteristic of the field of environmental education as a whole, it is the rapid succession of 'programs' and 'frameworks' that all pretend to break with former practices in order to make room for a new educational practice. Thus, the succession of ideas on what environmental education is for and about, clearly reflects the evolution of leading institutions and public opinion in society at large. Against this background, it is important to examine the history of environmental education – including its predecessors and closely related practices – in order to gain in-sight as to how environmental education became to be what it is at present. The brief history that follows is not complete, and comprises only a loose sketch of my view on the genealogy of environmental education. As such, the overview is only meant to provide a general background for the formulation of my research questions.

1.1 A BRIEF HISTORY OF ENVIRONMENTAL EDUCATION

Ever since human beings began to regard their relationship with the natural world as problematic in some way, they have presented this relationship as some kind of educational assignment. Consider, for instance, the idealistic educational program of the youth movement, emerging throughout Western Europe at the beginning of the twentieth century. Inspired as they were by the spirit of naturalistic romanticism, the scouting groups and *Wandervögeln* organised hiking trips into the woods, mountains and countryside to experience the wholesomeness of natural life. Obviously, this spirit was articulated much earlier by naturalist educational thinkers, such as Rousseau and Froebel, who argued that youthful innocence ought to be shielded from the distorting influences of modern society, so that they would learn to recognise nature as their 'hale arbiter' and follow their subsequent natural destination. But as the industrial revolution and urbanisation gave rise to an increasingly rapid expansion of industrial towns, educators felt more and more uncomfortable with the 'depravation' and 'alienation from nature' they saw as inherent to modern city life. Moreover, the general state of the cities and their urban poor became a matter of social concern. In order to retrieve a 'sound relationship with nature', nature education was introduced in most elementary schools. This introduction was partly due to the powerful lobby of the educational reform movement, that in similar fashion sought to overcome the alienation from the world inherent to the disciplined 'listening school'. Instead, this movement was ready to welcome 'full life' into the classroom. More specifically, nature education was aimed at teaching children about the 'amazing secrets of natural life'. Natural educators were convinced that studying nature – birds, bees, plants and trees – both inside as well as outside of the classroom, would increase love and respect for the natural environment. For them, knowing about and caring for nature was considered fundamental to the same attitude that was to be cultivated in school[2].

For a long period of time, nature education of this classical style continued to represent an important component of school curricula in Western Europe. However, under the influence of the emergent paradigm of scientific and technological progress, school curricula in general acquired a more utilitarian, positivist, technorationalistic character. Whereas nature education in elementary schools largely retained its traditional character – highlighting knowledge through acquaintance and life–world experience – this ideological shift implied that the secondary school curriculum gave privileged status to an ever more narrowly defined approach of natural phenomena in terms of scientific analysis. Within such a curriculum 'nature' appears mainly as an object of scientific description, explanation, prediction and effective utilisation for technological means, rather than as a source of admiration, contemplation, personal meaning, narrative and cultural significance – responsive dimensions of knowledge that were so directly evoked by the classical type of nature education. Thus, the intuitive unity of knowledge and care for nature was pushed to the rear, in favour of a more positivist approach to natural phenomena, positing quite an instrumental relationship to nature (cf. Bonnett, 2003, p. 646; Lijmbach, 2004, p. 8).

Halfway through the twentieth century, problems of urbanisation and pollution began threatening the integrity of the countryside and natural resources – forests, moorlands, wetlands, riverbanks and watersheds – in large parts of Western Europe and North America. Public awareness of these problems of nature conservation elicited strong educational responses; in the late fifties, the conservation movement successfully lobbied for the introduction of Conservation Education as part of geography courses taught at secondary and higher educational level, particularly in the English speaking world. The aim of conservation education was to connect children to their land and natural environment, not merely by means of knowledge that would convince them of the need for protection, but also by means of excursions into nature, so that pupils and students were able to experience their involvement with the natural environment as a source of aesthetic enjoyment and an 'ever-present power of recuperation' (Marsden, 1997). In the wake of Conservation Education, new educational initiatives found their way into the classroom, under such names as 'Nature Study' and 'Outdoor Education'.

Haunted by the alarming future scenarios of scientists and environmentalists in general, and the Club of Rome (1972) in particular, a growing awareness of the magnitude, seriousness and multidimensional nature of the environmental crisis emerged at the beginning of the nineteen seventies. Apart from the local problems of conservation, global problems of resource depletion in combination with population growth and world hunger, problems of climate change, the extinction of species, the unsafe storage of nuclear waste, acidification, ozone layer depletion, dehydration, the pollution of air, water and soil and natural degradation all started to attract public attention, albeit among a marginal group of progressive minds. Although scientists and environmentalists did not agree on all these problems in detail, their common message seemed to be clear: if we do not radically change our consumer behaviour, common practices and institutions within a short period of time, these problems will be aggravated and could ultimately lead to an irreversible situation. Newly founded groups of environmental activists, traditional conservation groups, student groups and associations of alarmed citizens all acted to mobilise social resistance against those industries, companies and governments that were responsible for these newly identified environmental problems. At the same time, environmentalists translated public indignation into political claims thus striving to put environmental issues on the political agenda. Their analysis of the environmental crisis was radical in the sense that they criticised the capitalist organisation of our global market economy and questioned the allied ideals of economic growth and technological progress for their persistent disregard of its burdens on our natural *habitat* and depletion of natural resources. Moreover, the environmental movement also informed the public about their view on the environmental crisis and started small-scale experiments with solar energy and biological farming in order to raise public awareness and support for possible alternatives (Waks, 1996). Thus, an international community emerged around the educational initiatives of this movement now predominantly labelled 'environmental education' and 'ecological education'.

Of great influence during this period was the international field work of an early founder, Bill Stapp, who defined the aims of environmental education as 'producing a citizenry that is knowledgeable concerning the biospherical environment and its associated problems, aware of how to help solve those problems, and motivated to work towards their solution' (Stapp, 1969, p. 30). In the same spirit of high-flown idealism, the goal of environmental education was defined by the so-called *Belgrade Charter* at the International Workshop on Environmental Education (1975): 'The goal of environmental education is: To develop a world population that is aware of, and concerned about, the environment and its associated problems, and which has the knowledge, skills, attitudes, motivations and commitment to work individually and collectively toward solutions of current problems and the prevention of new ones' (cited in: McKeown and Hopkins, 2003).

However, as the *International Encyclopaedia of Education* (1994) correctly mentions, environmental education was of greater interest to individuals and collectives outside of the formal education community – environmental activists, conservationists, natural resource managers, economists, sociologists, anthropologists and political leaders – than to the formal education establishment itself (Husén and Postlethwaite, 1994, p. 1991). This is evident from the fact that it took a long time before issues of environmental education were formally included in national curricula. One of the characteristics that made environmental education a *Fremdkörper* among the traditional contents of the formal curriculum, was its instrumental justification and subsequent purpose: as the previous quotes demonstrate, the practice of environmental education was not primarily intended to contribute to the actualisation of pupil's developmental potentialities, but designed to change the pupil's behaviour, attitude and mentality in a particular preconceived way, in order to create an environmentally sound world. In keeping with this advocacy, the aims of environmental education were almost exclusively defined in terms of individual dispositions, ranging from more general goals, such as 'creating environmental awareness', to rather specific objectives, such as the cultivation of environmental 'do's' and 'don'ts', persuading (future) citizens and consumers to reduce their use of energy, to pay attention to the kinds of products they buy, to travel by bus or train instead of by car, and so on.

More precisely, environmental education was supposed to contribute to the development of responsible environmental behaviour, mainly for the larger purpose of enhancing of the quality of human life, either stated in survivalist or humanitarian terms. More radical and holistic versions of environmental education opted for the protection and maintenance of the global biosphere 'for its own sake'. However, despite the survivalist, humanitarian or holistic language used to express these proposals, most initiatives in the field of environmental education insisted on a cognitivist presumption, namely, the presumption that cognitive understanding and awareness of the environmental problems that threaten our world will result in personal concern for the natural environment and responsible action on its behalf. Simultaneously, many educators and researchers at that time bemoaned the fact that the relationship between cognitive knowledge and behavioural change was not as simple as they first assumed it would be (Husén and Postlethwaite, 1994, p. 1992).

Despite the new rhetoric of the nineteen seventies, Lucie Sauvé argues that many initiatives in the field of environmental education retained the spirit of naturalistic romanticism and the characteristics of nature education. Almost simultaneously, however, she observes the gradual emergence of a grass roots environmental education movement that was more problem-oriented, socially critical and politically engaged in local community action (Sauvé, 1998). In the Netherlands, the field of tension between both 'spirits' manifested itself in an ideological conflict between proponents of 'green' nature education – focussing on personal experience and care for the natural life–world – and those in favour of a 'grey' environmental education – focussing on raising understanding and awareness of the structures of society that have caused and preserved environmental degradation, in order to foster the creation of a critical citizenry, prepared to reflect on their own consumer behaviour, and motivated to engage in collective action. This tension is discernible in other countries as well, and was mainly a conflict between the old-style conservation movement and the more radical environmental movement. In the Netherlands, this ideological debate eventually diminished into a kind of consensus that 'nature-and-environmental-education' (*natuur- en milieu-educatie*) – as it was called – should aim at both raising natural care and social awareness. However, in practice the schism between both orientations remained visible for a long time (Meijer, 1996; Praamsma, 1993, 1997, pp. 19–21, Verschoor, 1997).

In the early nineteen eighties, the institutionalisation of organisations and networks for environmental education increased, and gradually, environmental education itself became a slightly more mainstream phenomenon, although it took until the second half of the nineteen eighties before some national governments – among whom the Dutch – took the initiative to first fund and then formalise some of the educational practices. Furthermore, contacts with the formal school systems were also being strengthened.

A major ideological turning point in the history of environmental policy and education was marked by the presentation of the United Nations'-report *Our Common Future* (1987) by the World Commission on Environment and Development (WCED) under the supervision of Gro Harlem Brundtland, at the UN Conference in Stockholm. In this influential report, an extensive analysis is given of the environmental problems that arise due to the tension between the finiteness of the earth's natural resources and the infinite growth of human population and consumption. Thus, global environmental issues were analysed in connection with issues of (under)development and economic growth. Brundtland and her colleagues stressed that both poverty in the southern part of the world as well as excessive and indiscriminate economic growth in the northern part of the world have contributed to global environmental damage (*Our Common Future*, 1991, p. 28). This report was of immense importance, first because it intitiated a dialogue between environmental activists, conservationists, scientists and green politicians and representatives of international business and trade – two worlds that until then had failed to interact. It provided a common language that made this dialogue possible, namely, the language of sustainable development. The now familiar concept of sustainable development was introduced in this report and defined as 'a development that meets the needs of the present without compromising the ability of future generations to meet their own needs' (WCED, 1987, p. 12). As such, the concept was meant to stress the needs of those who are neglected by the present economic and political structures: the needs of people living in underdeveloped countries and the needs of future generations. Furthermore, the concept underlines that consideration for those needs, together with consideration for the earth's carrying capacity, sets limits on our present consumption and exploitation of the earth's natural resources. The harmful consequences of our actions for the well being of people in other places and in the future can no longer be 'externalised' from the calculations that inform local choice on the use of natural resources. The report went on to insist that social and environmental costs should be weighed by all those 'stakeholders'participating in the pursuit of a sustainable development.

Second, the Brundtland report was of major importance, because it included the first effort to globally unite institutional entities to lay the foundations for a common understanding and approach to the environmental crisis. Ever since 1987 and throughout the world, sustainable development has been regarded as the overarching aim of developmental and environmental policies – including environmental education. All UN countries that ratified this treaty are committed to designing their national policies within the framework of sustainable development. It is not surprising then, that the term is defined in quite general and abstract terms. Moreover, the definition of sustainable development was meant to have a 'second-order character', in order to leave open different local 'interpretations'. The demands of sustainable development should be adapted to local needs and circumstances, as articulated in the slogan: 'Think Global, Act Local'. However, due to its consensual character, the language of sustainable development is notoriously ambiguous. As we will see, the ideal of sustainable development pays lip service to those who are committed to sustaining natural integrity and environmental quality as well as to those who are mainly concerned with sustaining economic growth.

Michael Bonnett points to the seductive attractiveness of an appeal to sustainable development: 'it brings into harmony two highly attractive but potentially conflicting notions. First, is the idea of conserving or preserving those aspects of nature that are currently endangered through depletion, pollution and so forth. Second, is the idea of accommodating ongoing human aspirations to develop, that is, in some sense, to have more or better, where this necessarily has implications for natural systems' (cf. Bonnett, 2003, p. 676). Furthermore, the reports on sustainable development leave open so many conflicting interpretations on the possibility of reconciling the ideals of (environmental) sustainability and (economic) development within a global capitalist free market economy that it is hard if not impossible to disagree with its general analysis and guiding principles. It therefore comprises a lucky bag of good intentions, everyone can find something in the rhetoric of sustainability that suits her taste or conviction – radical environmentalist as well as well as modern captains of industry (cf. Bonnett, 2003, p. 681).

Unfortunately, this wide appeal has been reached at the expense of clarity and practicality. At face value, an appeal to the needs of future generations might appear helpful and concrete in practice, but as soon as 'stakeholders' refer to future needs, they either tend to veil present interests in terms of future needs or introduce a highly metaphysical, almost religious, authoritative argument that cannot be regarded a proper subject in public discourse. This is because it is not at all clear how the needs of future generations can be judged by us presentday contemporaries – as to how we should weigh their needs against ours, and how many future generations we should consider in our present deliberations. The idea of a sustainable development provides few indications as how to settle these questions. As soon as the guiding principles are applied in local practices, ideological tensions and oppositions between the different parties emerge when it comes to applying them, whereas the principle of sustainable development was supposed to bridge these ideological gaps.

It is important to dwell on the language of sustainable development, because ever since its major concepts and principles were introduced in 1987, the language has been of growing importance within the field of environmental education. The Brundtland report itself suggested only in general terms that 'Environmental education should be included in and should run throughout the other disciplines of the formal education curriculum at all levels – to foster a sense of responsibility for the state of the environment and to teach students how to monitor, protect and improve it. These objectives cannot be achieved without the involvement of students in the movement for a better environment, through such things as nature clubs and special interest groups' (*Our Common Future*, 1987, p. 113). Illustrative of the practical ways in which the ideal of sustainable development inspired new educational initiatives, are the lessons in which pupils learned to determine their *ecological footprint*. After listing their daily consumption of food and consumer goods, their means of transport, housing and use of services, pupils measure the total amount of productive land required to produce these resources and assimilate its waste products, expressed in amount of hectares (the ecological footprint of the average Canadian for instance adds up to 4.8 hectares). Thus, the ecological footprint was presented – and is still used – as a measure of how sustainable our life-styles are[3]. Obviously, the purpose is to stimulate pupils' reflection on the personal claims they lay on the earth's natural resources, and change their ways in such a manner as to fit within the limits of a sound ecological footprint, determined by the limits of the world's carrying capacity. This approach is a clear example of how environmental problems are presented not only as a matter of collective responsibility and global institutional reform but as a matter of individual consumer behaviour as well. Moreover, the idea emerged that the success of 'implementation' of sustainable development in education can be measured in terms of personal levels of consumption.

Whereas environmental education and sustainable development were only loosely connected in the Brundtland report, at the Earth Summit in Rio (1992) participating countries reached agreement over a global agenda for future action – the so called *Agenda 21* – in which environmental education is more narrowly defined as a 'tool' to advance sustainability. Chapter 36 of this agenda deals with issues of education and public awareness: 'Education is critical for promoting sustainable development and improving the capacity of the people to address environmental and developmental education, the latter needs to be incorporated as an essential part of learning. Both formal and non-formal education are indispensable to changing people's attitudes so that they have the capacity to assess and address their sustainable development concerns. It is also critical for achieving environmental and ethical awareness, values and attitudes, skills and behaviour, consistent with sustainable development and for effective public participation in decision-making' (*Agenda 21*, chapter 36.3). Again, the initiatives are regarded as successful or effective insofar as they bring about a change of 'unsustainable' patterns of production and consumption into more 'sustainable' ones.

In *Agenda 21* there is a key focus on informing consumers about the social and environmental impact of the products they buy, so that they are able to make 'environmentally informed choices'. There are three types of policy instruments at the disposal of national governments to promote sustainable patterns of consumption and lifestyles. First, there are legal instruments, simply obliging concern for the environment and prohibiting unsustainable behaviour, such as pollution. Second, there are economic instruments – environmental charges and taxes, deposit/refund systems – to influence consumer behaviour. Third there are the social instruments, which include education and public awareness programs that governments employ to impress on (future) citizens their environmental responsibility. 'Governments and private-sector organisations should promote more positive attitudes towards sustainable consumption through education, public awareness programs and other means, such as positive advertising of products and services that utilise environmentally sound technologies or encourage sustainable production and consumption patterns' (*Agenda 21*, 1992, chapter 4.26).

This strategy of a consumer-oriented education can be exemplified by the well-known project of *Eco-teams*, initiated after the Rio Summit but still strongly operative in many countries and at present modified for implementation within schools as part of ESD. Eco-teams is an educational programme, designed on the principles of behavioural change techniques to support households in their efforts to reduce their use of water, energy and waste on a grass-roots neighbourhood level. Eco-team participants – generally made up of six to eight households in a particular neighbourhood – gather every now and then to reflect on their consumption patterns, to develop and revise a plan of action as to how to reduce their use of natural resources and to present the measured results, and evaluate, exchange and advise one another[4].

In the course of the late nineteen nineties, new 'frameworks' and programs for environmental education were launched by local and national governments, environmental groups and institutions dedicated to the service of a general policy of sustainability. These programs were labelled *Education for Sustainable Development* (ESD), *Education for Sustainability (ES), Education for the Development of Sustainable Societies and Global Responsibility, Education for a Sustainable Future* (ESF), *Education for Our Common Future*, or simply *Sustainability Education* (SE)[5] (Sauvé, 1998, p. 44; cf. McKeown and Hopkins, 2003). As the titles themselves express, these educational programs were originally designed to promote 'sustainable consumption patterns and lifestyles' and strengthen public support for the agenda of sustainable development. As the *ESD Toolkit* mentions: 'Education is an essential tool for achieving sustainability' (McKeown, 2002). Thus, education is defined as 'the management of acquisition of knowledge, skills, attitudes and values by individuals to enable and empower them for sustainable practices, initiatives and participation as citizens' *(Education for Sustainable Development in Europe: 1997–1999)*[6]. This is underlined by the report of the United Nations Commission on Sustainable Development in 2001:

> *Agenda 21 recognizes education in all its forms (including public awareness and training) as an essential means for achieving progress towards sustainable development and for the implementation of all chapters of the Agenda. Education is no longer seen as an end in itself but rather as a key-instrument for bringing about the changes in knowledge, values, behaviour and lifestyles required to achieve sustainability'* (Education and public awareness for sustainable development, 2001, p. 2).

In response to this new approach of environmental education as an instrument of behavioural change, in the service of a policy of sustainable development, educational researchers criticised its presumed instrumental foundation, manipulative nature and normative content. This criticism was generally inspired by the liberal idea that education should not be employed in the service of some particular ideology. According to Bob Jickling – well known for his paper *Why I Don't Want My Children To Be Educated for Sustainable Development* (1994) – the real problem of these new frameworks for environmental education lies in the conceptual construction 'education for …'. In his view, education should not aim for something external to itself. Educators should not be invited to prescribe a preferred social end or way of life, but they should support the development of 'personal autonomy', 'critical thinking' and 'independent judgement'. These educational aims are assumed to be formal in character; they do not obtain a particular normative end, but provide the necessary competencies for children to choose their own ends. In this view, educators and pupils should always take a reflexive stance towards their subject. If environmental values of a particular ideological kind are firmly imposed on pupils and students, this may preclude the development of rational and moral autonomy. In particular *liberal* philosophers of education disapprove of such a strong commitment to predetermined values in environmental education because they fear that risks of manipulation, moralism, or even indoctrination lie in wait. They suggest that education should be *about* sustainable development, but not *for* anything, except for the education of autonomous and critical citizens (Jickling, 1991, 1993, 1994, 1997, 2001; Praamsma, 1993, 1997; Meijer, 1996, 1997; Jickling and Spork, 1998; Jans and Wildemeersch, 1999; Lijmbach, 2000, pp. 48–53).

Critics also responded to the external claims made on educational institutions and the heavy burden placed on the educational potential to change society. These critics stressed that, sometimes, educators should resist external claims to contribute to a predetermined approach of social problems. By virtue of its own task, purpose and position in society, the field of education should enjoy a certain amount of autonomy against the political and economic institutions in society. Educational institutions should not sacrifice their primary responsibility and expertise to thoughtless expectations from the outside. Some critics for instance argue that educators should not simply adopt the scientific and technological definitions of environmental problems, presented to them by environmentalists, policy-makers and scientists, because these definitions fail to connect to the life–world experience and environmental awareness of their pupils (Bolscho, 1998; Margandant, 1998). Moreover, educators are called on to translate social problems in such a way, they connect to the purpose and meaning of educational practice. Besides, critics warn of too high expectations for social change by pointing out those educational practices take place within the same world in which the social problems occur. Since these practices share in the identified social problems as well, education cannot just 'exclude' or 'remove' the inequality, injustice and environmental neglect from our world. By giving credence to the well-known slogan that 'education cannot compensate for society', these critics do not shake off the educator's social responsibility. Moreover, they underline their specific task in seeking answers to social problems (Meijer, 1996, 1997; Praamsma, 1997).

Furthermore, criticism was raised about the anthropocentrism inherent in the concept of sustainable development. After all, only the needs of (present and future) humans are considered in its definition. But even more than the Brundtland report – leaving open the possibility of an intrinsic appreciation of nature – *Agenda 21* (1992) is about protecting nature on earth for human's sake: nature as a human resource and a condition for human survival (Jickling, 1994; Bolscho, 1998). In line with this criticism, some critics discovered an allegiance to the neo-liberal agenda of sustaining economic growth and extending the free market economy within the language of sustainable development. Shiva suggests that in the present global context, the central concept of 'development' can hardly escape connotations derived from the capitalist market economy. Thus situated, economic development is immediately interpreted as economic growth, which in its turn is equivalent to the maximisation of profits and capital accumulation (cited from Bonnett, 2003, p. 680; cf. Bell, 2004, p. 46). This interpretation is supported by the 'twelfth principle' of sustainability as summarised in the *Rio Declaration on Environment and Development* (1992) and rehearsed in the *ESD Toolkit* (2002): 'Nations should cooperate to promote an open international economic system that will lead to economic growth and sustainable development in all countries. Environmental policies should not be used as an unjustifiable means of restricting international trade' (McKeown, 2002, p. 9). A similar UNESCO document suggests that ESD must promote 'creative and effective use of human potential and all forms of capital to ensure rapid and more equitable economic growth, with minimal impact on the environment" (UNESCO, 1992, p. 3). Here, education is first and foremost perceived as a 'central economic investment for the development of creativity, productivity, and competitiveness' (UNESCO, 1992, p. 14). It is not surprising then, that some critics see the concept of sustainable development as 'nothing more (or less) than a neo-colonial concept riding the waves of globalisation' (cited from: Hesselink et al., 2000, p.15). To be more precise, one could argue that the global striving for a sustainable development encompasses a recent expression of the neo-colonial strategy to subject third world countries to the rules of our global 'free' market economy, thus creating new potential markets of consumers and cheap labourers, and neutralising political turmoil. Furthermore, the ideal of sustainable development is used to conserve nature as a pool of resources to be utilised for a sustained economic growth. According to Sauvé, the main problem is that within the conceptual framework of sustainable development, the 'economic sphere' is viewed as a separate autonomous entity (next to the interlinked spheres of 'society' and 'environment': this triangle is sometimes expressed in terms of people, planet and profit) (Sauvé, 1998, 2002).

Previous criticism about the close connection between environmental education and sustainable development did inspire the proponents of these new frameworks and programs to reconsider the nature of environmental education and the status of sustainable development. That is, many participants in the international debate on environmental education responded to this criticism (cf. Hesselink et al., 2000; McKeown and Hopkins, 2003). However, judged from the outside, educational programs – especially those within formal school education – were framed even more strictly within the language of sustainable development. Gradually, ESD became the leading framework, in such a dominant way that other terms are simply ignored or neglected. Those who still choose to speak of 'environmental education' or 'nature education' are either corrected – ESD is the right term – or they are associated with an old-fashioned educational practice. In many contexts the term 'environment' or 'environmental' is being replaced by 'sustainability' and 'sustainable development'. A discussion about environmental problems beyond the framework of ESD is scarcely possible. Apart from a handful of educational researchers who persist in their criticism of the framework of sustainable development, the mainstream discussion on issues of environmental education takes place within this framework. The dominance of the language of 'sustainability' has recently been underlined by the UN Commission in Education and Communication, declaring the next decade as 'the United Nations Decade on Education for Sustainable Development 2005–2015' (IUCN, 2004). Meanwhile, ESD has become an established part of the curriculum in most European and many Third World nations, not as a separate discipline, but as a cross-curricular theme or field of learning (Bonnett, 2003, p. 675).

1.2 ESD: RESEARCH QUESTIONS AND DIRECTIONS

The fundamental questions guiding my research are informed by the present debate on the relationship between 'education' and 'sustainable development'. Within this debate, two views can be discerned. First, there is the *positivist view*[7], dominant in circles of environmental management and policy-making, holding that education should be mobilised as a vehicle for the promotion of sustainable knowledge, preferences, attitudes and patterns of behaviour that reflect the requirements of sustainable development. According to Elliot, this view is related to the school effectiveness movement, because it prespecifies generalised tangible outcomes (referred to by Bonnett, 2003, p. 10). Informed by subject experts – mainly natural scientists – it is therefore assumed that environmental educators know what sustainability requires and that they impose the required knowledge, values, attitudes and behaviour patterns onto those who are assumed to be ignorant on this point, namely the pupils or students. Thus, ESD is seen as a rather systematic, technical enterprise of transmission, the success of which can be measured in terms of desired consumption levels. Furthermore, as Bonnett outlines, it is tacitly assumed that 'its underlying values are largely economic and unproblematic', and that 'the implications of sustainable development for the moral/social/political structure of society are basically consistent with the status quo. Understood in this way 'sustainable development' rapidly converges with 'common sense' and an instrumental rationality determines the means for achieving a set of taken-for-granted ends' (Bonnett, 2003, p. 10).

Opposed to this positivist view of ESD is the *constructivist view*, dominant in educational circles, assuming that schools ought to further sustainable development by encouraging ongoing pupil exploration and engagement with environmental issues. Here the essence is to develop pupils own critical ability and interpretations of issues in the context of firsthand practical situations that they confront. This 'action competence approach' is more consistent with the 'school development movement' (Bonnett, 2003, p. 10). Rather than adhering to a firm ideology or fixed end, proponents of this view see sustainable development as an 'agenda' or 'conceptual vocabulary' which makes discussion possible among the disparate stakeholders of sustainable development in local practices. The content and requirements of sustainable development are never fixed, but locally defined by the consumers, producers, politicians, environmentalists and other participants who deal with local problems in search of a common ground. Thus, the effort to achieve sustainable development can be characterised as a continuous dialogue and social learning process, in which social-cultural, ecological and economic interests are weighed against the background of particular local practices and problems. There is no preconceived balance between the interests of 'people, planet and profit', but this balance is constructed through 'learning by doing and doing by learning'. In such a process of social learning, pupils are now appealed to as consumers or holiday makers, then as family-members, traffic participants, employees or neighbours, so that they learn to cope with various roles and interests (Hesselink et al., 2000; cf. Stuurgroep Leren voor Duurzaamheid, 2003).

The questions guiding my research concern two fundamental controversies between the positivist and constructivist view on ESD. First of all, there is the issue of intergenerational citizenship, secondly, the issue of natural value and care. To start with the first: inherent to the framework of ESD is the imperative idea that we should imagine ourselves as citizens of a world community stretching out over several generations with whom we are to share the natural resources that are conditional on our survival and well-being. This ideal of intergenerational citizenship requires us to reconsider our present use and exploitation of natural resources, measured against the standards provided by hypothetical claims of future people; those unknown people with whom we do not share and will not share a common life. Differences between the positivists and constructivist view on ESD are partly due to a disagreement on the question as to how citizens are to consider the hypothetical needs and interests of future generations. According to the positivist view it is possible to prespecify the basic needs of future people and delineate the demands on our patterns of consumption and production in terms of measurable outcomes. Thus, the exact limits of our 'ecological footprints' can be settled in accordance with a distributive principle of justice. These requirements on our behaviour – which will be minimal in effect – should be transmitted onto new generations of citizens within educational settings, so that they learn to justify their behaviour against these settled standards and comply with the sustainable codes of conduct. Here, education is regarded as a mere instrument, employed to achieve behaviour change in accordance with settled, external ends. In this view, the determination of educational ends is the prerogative of natural scientists and experts (Bell, 2004; Bonnett, 2003, p. 699).

Whereas the positivist view relies on a definition of sustainable development provided by a regime of scientists and experts, the constructivist view on ESD relies on local practices of deliberation, as previously outlined. In this view, the content and requirements of sustainable development are to be defined by local stakeholders, participating in a collective search for local answers to environmental problems. As such, sustainable development is part of a learning process, in a broad social as well as an educational sense. Rather than a complex of passive entitlements and some heavily circumscribed duties, citizenship implies an active responsibility for common affairs and requires us to participate in public life. However, what is unclear are the means the constructivists use to judge the hypothetical claims of future generations. Are these claims only to be judged as general appeals to our future responsibility, or does the agenda of sustainable development still prespecify particular outcomes, albeit in more general terms?

Thus, a sharp opposition becomes evident between the positivist view – prescribing an ultimately minimal, though substantive morality of sustainable development – and the constructivist view – offering a broad agenda, though without responsibilities. Within the liberal framework of ESD, debates on the nature of environmental education are generally trapped in an ideal–typical opposition between two extreme practices. One extreme leads to a wishy-washy domestication of environmentalism – treating ecological values and standards as mere lifestyle options – the other ends in practices of indoctrination – imposing sustainable development as a firm ideology upon (future) citizens. In my view, this unfruitful trap is due to the fact that both views take the well being, needs and rights of future generations as the primary object of our responsibility as citizens. In the next chapter, these intricacies of an intergenerational understanding of the political community will be explored, and the outlines of an alternative, neo-republican understanding of future responsibility and environmental citizenship will be developed.

Positivists and constructivists within the field of ESD furthermore disagree on the issue of natural value and care. At face value, the framework of ESD appears to be of a rather anthropocentric nature. The principle of sustainable development requires us to reflect on our involvement with the natural environment in economic terms. After all, the language of sustainable development is about 'human resources', 'economic development' and 'satisfying needs'. Thus, our relationship with nature is narrowly understood as a relation between consumer and commodity. The fundamental awareness that nature might be valuable and good for its own sake, independent of our use, consumption and exploitation – an awareness of intrinsic natural value – is excluded from the discourse on the environmental crisis as an economic problem of resource-management and distribution. However, by taking sustainable development as the predetermined end of ESD, only the positivist view commits itself to the literal definition of sustainable development, as outlined in the Brundtland report – *Our Common Future* (1987) – and *Agenda 21* (1992). Many proponents of the constructivist view are not comfortable with this anthropocentric reading of the framework of ESD. And indeed, due to its consensual nature, there are indications in official (E)SD-documents that allow for a less narrow economic interpretation, leaving room for considerations of intrinsic natural value. According to some constructivists, the framework of ESD is compatible with an ecocentric approach of the environmental crisis. Others even argue that ESD is more 'critical of the predominant market and consumption driven society' and 'more open to new ways of thinking and doing' than the old-style environmental education (Hesselink et al. 2000, p. 15).

In order to distance themselves from the narrow anthropocentric interpretations, some constructivists deliberately avoid the term sustainable development, and choose to speak of 'education for a sustainable future', 'education for a sustainable society' or simply 'education for sustainability'. Definitions of these frameworks of sustainability are similar but slightly different from the definition of ESD. However, it is doubtful whether these dissidents in the ESD camp will be able to maintain their alternative understanding of sustainability against a vast majority of readers who will immediately associate sustainability with 'sustainable development' and its most influential expressions in Brundtland's *Our Common Future* (1987) and *Agenda 21* (1992).

Behind this word play hides the fundamental issue whether we humans are able to escape an instrumental valuation of nature. If one insists that the value of nature is intrinsic to nature, like constructivists assume, how can human evaluators experience this value without 'using' nature somehow as a resource for meaning or source of appropriation? And if one argues, like some positivists do, that natural value is no more than a label attributed by human evaluators, how can this value be seen as intrinsic to nature? ESD debates on the intrinsic value of nature are anything but clear on this point. In itself this vagueness would not be a major problem, if it were not of fundamental importance for our understanding of environmental responsibility. Intrinsic value claims can be understood as expressions of responsibility; by expressing the things in life we really care for, we commit ourselves to preservation of its meaning and pursuit of its goodness

What positivist and constructivist views on ESD have in common is their tendency to look for an answer to the issue of intrinsicallity and responsibility somewhere in the opposition of nature and its human evaluators, that is, between human subject and natural object. In chapter three, more detailed meta-ethical study will show that both subjectivist and objectivist accounts of natural value ignore the social nature of valuation. The judgements in which we assume responsibility for our natural environment are part of intersubjective practices of care for our natural environment. In line with this insight, I will argue that environmental education requires initiation into collective practices within which our caring involvement with the natural environment is preserved. Children gradually assume responsibility, not by practicing skills of personal choice, but through involvement in collective deliberation and action. Furthermore, I will characterise the nature and scope of environmental responsibility in close connection to Merleau-Ponty's phenomenological study of humans' bodily involvement with the natural environment as displayed in our everyday perception and care for the things in our life–world.

These studies on the nature of future responsibility and responsibility for our natural life–world are not only meant to result in profound criticism on the framework of ESD. The conclusions are also meant to inform an alternative perspective on environmental education. In the last chapter this perspective will be presented, implying an alternative understanding of environmental sensibilility, responsibility and knowledge of environmental issues. Furthermore, I will illustrate my perspective in a proposal for an alternative framework and program for environmnetal education, to be called *education for environmental responsibility*.

The general aims underpinning this research are modest and mainly of a descriptive kind. My purpose is to describe the guiding concepts and conceptual relationships within ESD, to reveal the inherent tensions, contradictions and practical implications within its language and dwell on some of the underlying philosophical issues concerning future responsibility, intrinsic natural value and care. However, despite my commitment to 'descriptivism' in this broad sense, I do not think that any author can describe a social practice without simultaneously expressing a particular engagement with the practice described. By presenting a particular picture of reality, philosophers do open up new spaces and possibilities for action. Thus, the practical purpose of this philosophical investigation should be understood in this light. As my descriptive analysis proceeds, I hope to reveal the strengths and weaknesses of the present understanding of environmental education within the framework of ESD, while simultaneously suggesting possibilities for an alternative understanding. Particularly, in the section 3.4 of this book I will present an alternative framework for environmental education: an education for environmental responsibility.

This research is mainly dedicated to a philosophical enquiry into *the language* of environmental responsibility and education, conducted in line with the ordinary language philosophy, as presented by Wittgenstein in his *Philosophical Investigations* (1953), and exemplified by Peter Winch in *The Idea of a Social Science* (1958). However, when I speak of 'nature', 'future' or 'education' my concern is not limited to what people say and write about 'nature', 'future' or 'education'. Such a limitation would imply an unnecessary nominalist reduction of our experience of reality to linguistic experience. Indeed, my concern is with the nature that moves me, with the future that appeals to me or frightens me, and with the education that we practice and experience in everyday life. But these existential 'objects' of experience are only meaningful to us insofar as they are able to be articulated in a common language – not necessarily a language of spoken or written words, but more broadly understood as a common horizon of symbols and references, in which people express and share meanings with one another (such as in the various ways we also express ourselves using body language, facial expression and sign language). Therefore, an enquiry of the meaning of nature, future and education is necessarily an enquiry of the expression of nature, future and education in human behaviour and action. We cannot get behind these concepts, straight to 'the phenomena themselves', since our reference to these experiences is always mediated through language.

The language we are analysing is not taken to represent an external reality. Our concepts of 'nature', 'future' and 'education' do not correspond to the things itself 'out there in the real world'. Rather, language and reality are co-emergent (*gleichursprünglich*), as Wittgenstein articulates in his proposition that an agreement in judgement is necessarily also a judgement in form of life (Wittgenstein, 1958, proposition 241). This does not mean that there is nothing outside our language, but it does mean that the 'things out there' – the sheer existence of things and our reactive responses to them – have to be expressed in a common language in order to be significant, in order to have a meaning and place in our world. Without expression, the sheer thereness of things remains unintelligible and diffuse in such a way that it cannot be grasped in our mind. Subsequently, by expressing what 'nature', 'future' or 'education' means to us, we inescapably render ourselves susceptible to responsive claims of others, who make sense of what we say and do according to the rules of meaningful behaviour that operate within the particular practice in which we find ourselves.

I would like to discuss my understanding and use of the concept of nature to be more precise about the implications this epistemological framework will have for the aims of my research. While I develop my own tentative definition of nature in chapter three, such an exercise will necessarily be disappointing, since no definition of nature is adequate enough to cover its full and deep meaning and presence in human life. Any definition highlights only one particular aspect, or perhaps a few aspects, but the full meaning of nature remains elusive. Nature in this metaphysical sense apparently transcends our language, and as further analysis will show, our language transcends individual preferences. It is not necessary to go deeper into this 'double transcendence' here, since this complex issue will be discussed in more detail in chapter three. But by insisting on this transcendence I would like to stress that my rejection of the idea of representative language does not imply that any concept of nature is just a linguistic construction. On the contrary, there is always more to nature than we express in our words and display in our behaviour. There is, as it were, a transcendent surplus of meaning. Consequently, there is always more to nature than philosophers are able to reconstruct in terms of concepts and conceptual schemes. As any concept of nature, my concept of nature is an expression that is necessarily one-sided, but it is nevertheless capable of evoking a fuller meaning than allowed for in our present use of the concept. The transcendent surplus of meaning cannot be represented nor referred to in our concepts, but evocative language can remind us of this surplus. Not only the evocative language of the poet or philosopher, but ordinary language as well: when I speak of 'the sky', many connotations and personal meanings will resonate with the word that cannot be conceptually analysed such as meanings that have to do with the experience of infinity and human triviality, the aesthetic fulfilment we find in its beauty, personal memories of the nights when I have slept under a starry sky, and so on.

Just like our concept of time, our concept of nature might be labelled as a 'reminder' in the sense that Wittgenstein borrows from Augustine: 'something that we know when no one asks us, but no longer know when we are supposed to give an account of it [cf. Augustine], is something we need to remind ourselves of' (Wittgenstein, 1953, no. 89). This means that we can either embrace nature or speak about it. The words we use to express nature remind us of something, but this 'something' exceeds our ability to express it. In its inexhaustible fullness and elusive complexity the essential reality of nature remains unspeakable to us, and therefore, incommunicable. However, if we assume that there is nature beyond our language in this particular sense, what is the awkward status of this 'nature'? A tentative answer to this question can be found in Wittgenstein's notion of the 'form of life' as the inescapable horizon of certainties against the background of which we act. These are foundational certainties like our experience of gravity and the experience of colours, comprising the very substratum of our human existence. In our action we anticipate natural things contrafacticly. We do not dispute the existence of the particular tree that we find on our path from A to B, but we simply walk around the tree. If we nevertheless choose to express in words 'I walk around the tree' then this assertion should be understood as an expression of a foundational belief in our coexistence with the natural things that our senses present to us. Ergo, whereas we cannot legitimately deny nor confirm 'the reality of the tree' in a philosophical sense, at the level of everyday practical experience, we might insist that our behaviour in this particular situation would be completely insignificant unless we assume that there is a tree. In this sense, there might be a shared reference outside of language but we cannot articulate it. It is not the referent – nature itself – but the shared practice of referring to nature that functions as the ultimate warrant for the coherence of meaning of the word 'nature'.

So much for the social background, questions, epistemology and ambitions of my philosophical inquiry. Let us now turn to the heart of the inquiry itself: the issue of responsibility for our future world.

NOTES

[1] Orwell, 1948, p. 46.

[2] Apart from (translations of) the term 'nature education', these school disciplines were sometimes labelled differently, like the German 'Heimatskunde' and the Dutch 'heemkunde'.

[3] Check for instance the website: http://www.sustainabilityed.org/ef.htm

[4] http://www.ozgreen.org.au/lwlc/eco-team.htm

[5] Simultaneously, *Duurzaamheidseducatie (Sustainability Education)* was introduced in the Netherlands (followed by *Leren voor Duurzame Ontwikkeling [Learning for Sustainable Development]* en *Leren voor Duurzaamheid [Learning for Sustainability]*) and *Bildung für Nachhaltige Entwicklung [Education for Sustainable Development]* in Germany.

[6] http://www.hect.nl/hecteducationforsustainabledevelopmentineurope.html

[7] The terms 'positivist' and 'constructivist' view are drawn from Hesselink et al. (2000, p. 41). More or less similar distinctions are made by Bonnett, who chooses to label them respectively as the 'environmentalist' and 'democratic approach' of environmental education, and those who distinguish between 'weak' and 'strong sustainability' (cf. Bell, 2004, p. 42).

CHAPTER TWO

BECAUSE WE ARE CITIZENS

Exactly for the sake of what is new and revolutionary in every child, education must be conservative.

Hannah Arendt[1]

Despite all the controversy about what environmental education ought to be for and about, the participants in public discourse on environmental education would agree that it should prepare children for the practice of environmental citizenship. And despite what this practice implies, the participants would also agree that environmental citizenship involves responsibilities for presentday as well as future generations. They recognise the imperative of responding to future needs. We cannot remain indifferent to the demands that future people make on us, here and now, by virtue of the fact that they will inherit the world we leave behind for them. Our sense of justice tells us we ought to consider the implications and consequences of our present behaviour for the quality of future life on this globe. Just as our ancestors consciously contributed to the world we live in, we should contribute to a liveable world for posterity. This moral ideal corresponds closely to our common sense intuitions. In slogans and phrases used by political parties, for instance, future generations or 'our great-grand children' are often appealed to as an incentive for change. Furthermore, there is the old saying that 'we do not inherit the earth from our parents, we borrow it from our children'. And the biblical metaphor of 'stewardship' contains a clear notion of responsibility stretching out over several succeeding generations. But perhaps the most influential articulation of intergenerational responsibility is offered by the international key-concept of 'sustainable development', defining 'a development that meets the needs of the present without comprising the ability of future generations to meet their own needs' (*Our Common Future*, 1987).

When environmentalists, scientists and politicians speak on behalf of future generations they refer not only to our children, grandchildren and great-grandchildren, but all those unknown people who will live in the future, centuries from now. People about whose probable existence we can surmise, even though we cannot be sure. Obviously, the scope of our responsibility follows from the magnitude and widening radius of our actions. We can no longer stick to our initial responsibility for immediate descendants whom we care about, since, within the present context of rapid economic and scientific development, our personal consumption and use of technology will have an impact on the world and well-being of people who will live centuries from now. It is questionable, for instance, whether our present use of nuclear energy is consistent with an imperative of intergenerational responsibility, for, as long as scientists cannot guarantee safe and sustainable storage of nuclear waste, we consciously take huge risks for future generations who may suffer from leakage. The same can be argued with regard to the poisonous waste we produce and that will maintain its devastating effects on human health for thousands of years, or the uncertain effects of genetic engineering on the 'health' of our ecosystem in the long run. Whereas the consequences of human behaviour used to be confined to a surveyable distance in time and space, right now the intended and unintended effects of our behaviour stretch into the indefinite future with a magnitude that seems unprecedented. Therefore, we find ourselves in a position of increasing responsibility and uncertainty at the same time. We are called on to enter into a hypothetical dialogue with future strangers, since we have unintentionally entered their threshold and knocked on their door. Obviously, an adequate account of environmental citizenship has to respond to these changing conditions of human responsibility.

A shared commitment to an ethic of intergenerational responsibility seems to provide environmental educators with a strong incentive for environmental citizenship, but their common ground is weak and superficial, rather than firmly principled and circumscribed. As soon as we pose the question on what grounds and how we should consider future needs or include the 'voice' of future generations in our civil practices, we leave this common ground. There are many different views on the nature and ethical implications of our relationship with future generations. What kind of relationship with future people commits us to act in their behalf? What in this relationship inspires us to take up responsibility? What do these responsibilities consist of and how far do they reach? In this chapter I want to explore three possible views on this issue of intergenerational responsibility and citizenship.

In section 2.1 I will present the standard liberal view underlying the dominant understanding of the framework of sustainable development, that requires us to regard future generations as hypothetical fellow citizens. According to this view, we should represent the needs of future generations in public discourse by acting *as if* they were present. Or more specifically, we are supposed to judge and act in line with an intergenerational contract of distributive justice that can count on the hypothetical consent of future generations. Thus, our community of justice is being extended into the indefinite future, and the rights of contemporaries are, in principle, equally important as the conditional rights of future citizens. In this way liberal contractualists have a strong case for intergenerational justice. However, as I will argue, a principle of intergenerational justice cannot be justified within the liberal framework it presupposes because future generations are beyond the reach of our reciprocal relationships (in the narrow sense). Besides, it is unclear how such an abstract principle of intergenerational justice should motivate us to take up responsibility for the well-being of future generations.

My identification of fundamental flaws within the liberal view will lead us to explore the communitarian view on intergenerational responsibility in section 2.2. Communitarian ethicists suggest that we should treat future generations as heirs of our moral community. We are called on to contribute to their 'good' since future members of our community share our ideals and conceptions of the good life. After our death, they will continue the practices and projects we cared for and invested in during our life. Without trusting that future generations will carry on our traditions and pursue our ideals in the future, our life here and now would hardly be worth living. Thus, our relationship with future generations is understood and experienced in terms of extended community ties. Such an extension makes it possible to make obligations to posterity derived from the commitment and solidarity expected within the present community. Although the communitarian view offers an adequate answer to the problem of motivation, the responsibilities that it gives rise to remain limited to members of the community. Though, I will argue this view rests on an inadequate and outmoded picture of society.

In section 2.3, I will argue that the fundamental flaws of the liberal view, together with the shortcomings of the communitarian view force us to adopt an alternative understanding of future generations as imagined strangers. Although I draw on different sources, I will call this a neo-republican view of intergenerational responsibility. Within this view the main focus shifts from the alleged needs, wants and rights of future generations to our present desires, expectations and imagination of the future world that bind us together here and now. We do not know about people in the future, nor their needs and wants. Therefore, every judgement on our part as to what kind of world people in the future will like, what 'resources' they will need to survive and what 'goods' will make their life worthwhile, reflects what we like and what we think 'worthwhile' and 'good' in our present world. Every attempt to include the rights, needs or goods of future citizens throws us back upon our own devices as to what we value as 'necessary', 'liveable', 'inalienable' or 'dignified'. This does not mean that we do not have to consider the existence of future generations. On the contrary, we should imagine what kind of world we wish them to live in, rather than calculating what the interests of its inhabitants could possibly be.

As I will argue, this kind of future responsibility does not arise from moral lessons but will emerge in the process of public debate and participation in the civil society itself. Therefore, environmental education should consist of an introduction in the practices of this civil society, its debates on the environmental crisis and its collective efforts to conserve what we care for. In section 2.4 I will show that such an introduction and participation will ideally leave open our future to those who are new in this world. At the same time, guided participation will have to endow them with a basic trust in the indeterminate future. Such an open time horizon is of fundamental importance since it is requisite to the very possibility of education in general and of environmental education in particular.

Some preliminary remarks should be made before we turn to the philosophical investigation of future responsibility. First of all, one should be cognizant of the fact that the three perspectives on future responsibility and citizenship that will be distinguished in this chapter are expressions of ideas rather than practices. Liberalism, communitarianism and neo-republicanism represent more or less clearly distinguished schools of thought in contemporary political philosophy, but these positions are rarely if ever to be found in pure form within contemporary political practice. On the contrary, most practices of citizenship throughout the world bear traces of all three schools of thought. An acknowledgement of the fact that the basic system of rights and duties in most western countries reflects the liberal contract doctrine does not deny that the way in which communities express themselves alludes to the ideals of communitarianism, nor that the public culture of debate and protest is inspired by neo-republican thinking. These practices of citizenship coexist within post-modern society, nor are they mutually exclusive in all respects, even within political philosophy. There may be some issues on which the positions of these three perspectives diverge, but on other issues they converge and overlap. Moreover, liberalism, communitarianism and neo-republicanism are distinguished in terms of the paradigmatical issues they address and the emphasis they place on particular practices, rather than on their formal principles or framework of reference. While liberals ask themselves what a state could possibly demand of its citizens, and *vice versa*, what its citizens can demand of the state, communitarian critics are mainly concerned with the problem of how to assure the continued existence of their moral community. Neo-republican theorists, on the contrary, are interested in the question of how citizens act collectively in response to collective problems and thereby shape their lives, i.e. how they appeal to one another's responsibility in virtue of their being citizens.

When I use the terms 'liberalism' or 'liberal morality', I refer to the dominant framework of contractual liberalism, as articulated by John Rawls in his *Theory of Justice* (1971; and the revised edition of 1999) and *Political Liberalism* (1993). My discussion of the communitarian view on future responsibility focuses largely on the theory developed by Avner de-Shalit in his book *Why Posterity Matters* (1995) and some papers of early precursors like Martin Golding. My selection of these representatives is arbitrary but not coincidental; Rawls and de-Shalit were simply the authors who I frequently encountered in the current discourse on future responsibility and recognised as interesting and influential spokesmen. My use of the term 'neo-republicanism' has a slightly different background, primarily because there is not a well-defined theory of future responsibility available within this school of thought. In the second place, I chose to draw from the ideas and arguments of neo-republican thinkers like Hannah Arendt and advocates of a radical democracy like Chantal Mouffe in order to formulate my own perspective on the issue of future responsibility that copes with the weaknesses of the other two perspectives and aims to transcend the opposition between 'formal' liberalism and 'substantive' communitarianism.

My final remarks amount to the inadequacy of speaking in terms of *the* communitarian or *the* liberal position. Obviously, both schools of thought harbour many different classes and subclasses that allow for divergent perspectives on the issue of future responsibility that all carry the predicate 'liberal' or 'communitarian'. As will become clear, there are even different positions *within* the Rawlsian liberal framework. Partly due to the fact that Rawls expressed his idea in rather formal terms, many interpreters have applied his principles and insights in a practical way that moves beyond the formal setting and scope of his initial theory. As such, Rawls' theory of justice has engendered an interpretative discourse that might be at odds with his initial purposes in some respects. However, it is not my prime ambition to do justice to the purposes that Rawls had in mind when he defined his theory, but it is the Rawlsian framework that is my main object of concern. In my discussion of all three strands of thought, I will give internal as well as external criticism. Within the context of the latter, I will criticise both Rawls and de-Shalit in terms of the social practices to which their theoretical concepts give rise. For instance, I will argue that the application of the liberal distinction between private and public concerns is likely to privilege those consumerist lifestyles and conceptions of the good life that are consonant with the culture of free market exchange. Perhaps this implication is not in line with Rawls' purposes nor with the ideas of his interpreters, and obviously, it is not a logically necessary implication of the Rawlsian framework. However, arguments of a more contingent-empirical nature allow me to argue that the application of liberal concepts and principles in practice tend to work out in this or that particular way (that might indeed stand in sharp contrast to the purposes of their spokesmen). When it comes to this, my criticism is of a social rather than a logical kind. These preliminary remarks are largely reminiscent of the well-known Wittgensteinian point that a rule or principle does not carry within itself the means of its application. The application of a theory of responsibility in a particular intergenerational context brings along a particular cultural praxis, and together with this, all the contingencies that give meaning to who we think we are and what we should do in the face of an open and infinite future.

2.1 FUTURE GENERATIONS AS FELLOW CITIZENS

In the vast majority of democratic states throughout the modern world the relationship between citizen and state is governed by a so-called 'liberal contract', defining the basic rights and duties of both parties. These contracts start from the basic assumption that in modern pluralistic society there is no common view on what makes life worth living. Moreover, liberal thinkers argue that there is no rational way – no publicly specifiable principle or form of reasoning – that could determine the truth, goodness or beauty of one conception of the good life above the truth, goodness or beauty of the others. According to liberal theory, individual citizens should be able to make up their own mind in moral, ideological and religious matters. In line with the etymology of the word autonomy – comprised of *autos* (self) and *nomos* (law) – citizens should judge in accordance with a self-chosen moral law. This conviction leads liberal thinkers to argue in favour of a contract that guarantees the inalienable right of all citizens to develop their own conception of the good and to live according to their own life-plan, safeguarded from immediate state control or interference. Therefore, the liberal state in general, and the school as one of its institutions, should not promote a particular morality or conception of the good. Yet, the liberal state does not have to remain neutral to all moral dilemmas in society. In contrast to its neutrality regarding first-order conceptions of the good, the liberal state is morally bound to promote and protect liberal second-order values: those values, strictly necessary to guarantee democracy and personal freedom, such as tolerance, respect for diversity, non-violence and open-mindedness. Liberal morality, comprised of these second-order values, does not compete with the manifold (first-order) conceptions of the good, but supports this moral and ideological diversity. In fact, liberal morality is conditional on this diversity; it encompasses the necessary conditions for freedom of consciousness, thought and speech (cf. De Jong, 1998).

Of fundamental importance for the liberal understanding of personal autonomy, politics and public morality is the sharp distinction between the private and public sphere. As previously suggested, the private sphere is celebrated as the primary space where people are presumed to find ultimate life fulfilment by living according to their own device, taste, religion or view on life in 'the pursuit of happiness'. This 'conception of the good life' that people enjoy and practice within the private sphere is generally taken to imply substantive values, tastes and opinions on what makes our life worth living, identifications to the communities and movements we feel ourselves bound to, opinions on environmental issues, ideas on what the ultimate meaning of nature consists of and what kind of involvement with the natural environment we would like to pass on to our children. As such, the private sphere is the space that gives room for ideological diversity (within 'reasonable' assumed limits). In fact, the private sphere is understood as a pre-political space; the arrangements, relationships and values of family, neighbourhood, profession and community are seen as given, that is, beyond the scope of political interest. The only limitation to this freedom in the private sphere is imposed by the freedom of others. Ergo, in order to coördinate the maximisation of everyone's private freedom, some public arrangements have to be agreed upon and settled in a contract. The central institution that regulates and supervises the compliance with theses regulations is the state. Now, the public space is negatively defined as the non-private space, or put in positive terms, the space that is marked by the relationship between citizens and state. Political action and debate within this space is supposed to take place within the boundaries of 'public reason'. This means that citizens are required to express their interests and opinions in terms of arguments that do not originate from and are therefore independent of their particular conception of the good life. In principle those reasons should be acceptable by all citizens on 'reasonable grounds', regardless of their private background. This assumption of ideological independence of public and private sphere is of fundamental importance to safeguard state neutrality. Moreover, the belief in a non-ideological space is a constituent part of the standard-liberal understanding of state neutrality, that is, the state is required to justify its action and interventions without reference to an ideological or religious doctrine.

Environmental policy in general and environmental education in particular, *do* promote a particular morality. After all, with environmental education governments aim at a change of mentality and behaviour, in public as well as in private life; they seek to persuade citizens to reduce their use of energy, to pay attention to the kinds of products they buy, to travel by bus or train instead of by car, and so on. Therewith, environmental education distinguishes between more and less 'sustainable' lifestyles and conceptions of the good. Based on this normative distinction both state and school interfere with the choices people make within their private sphere. By doing this, governments inevitably value one conception of the good life above the other. These policies therefore seem to conflict with the normative requirements of liberal democracy *unless* the environmental values at hand can be justified within the framework of liberal morality itself; for then the promotion of environmental morality should, on the contrary, be regarded as a core responsibility of the liberal state, and hence, of liberal education. If, on the other hand, environmental values cannot be justified in a liberal way, then environmental morality has to be regarded as a private conception of the good, and, consequently, environmental education would represent an illegitimate form of state paternalism. So, according to liberal morality, the legitimacy of environmental education depends on the question whether its morality should be identified as a particular conception of the good life or as a necessary part of every liberal contract (cf. Bell, 2004, p. 39)[2].

Thus, the liberal state finds itself caught in a double bind. For, on the one hand, a normative form of environmental education seems necessary if the seriousness and urgency of environmental concerns are taken into account. On the other hand, however, normative interference seems undesirable, for reasons of autonomy. Among liberal philosophers there is a substantial debate on the range of environmental values that a liberal state should be permitted to promote actively and impose upon its citizens. To what extent can liberal morality be broadened in such a way, that it is able to include environmental values, necessary to justify a normative environmental policy and education? In order to achieve such a broadening of liberal morality, different authors on this subject start from different principles, and follow different routes from there. Within the context of this chapter it is not possible to give an extensive overview of all the approaches relevant to this ethical problem. Therefore, in the remaining part of this chapter the possibilities of three approaches, which are most promising, will be examined and scrutinised. These approaches depart from principles belonging to the canon of liberal ethics.

First, the basic intuition that 'thou shall not harm' can be applied to environmental concerns. After all, in the majority of cases, environmental pollution, degradation and unsustainable behaviour will eventually cause harm to human property, health or freedom. Second, the intuition represented by the primary goods principle implies that the liberal state ought to provide and protect those goods which are in everyone's interest in general and in nobody's interest in particular. Perhaps, a sound and sustainable environment can be regarded as in everyone's interest, without favouring one 'way of life' or 'conception of the good' over the other. A sound and sustainable environment seems preconditional on whatever life individual citizens want to lead. Third, we will see that the limitations of both the harm and primary goods principle force liberals to explore the possibility of a principle of intergenerational justice. If liberal morality is to cover our responsibility for the long term effects of environmentally harmful behaviour and the primary goods of future citizens, then its scope should reach beyond the moral community of those living at present. At this point, our conviction that future generations are entitled to a certain quality of life will be severely tested; for is this moral intuition strong enough to motivate and justify sacrifices which weigh heavily on our present welfare and freedom?

Environmental harm and risk

Liberal theory does not promote personal freedom *ad infinitum*. While the contract is said to allow all people to develop, pursue and revise their own conception of the good, it will not allow anyone to violate the freedom rights of others. Logically speaking, everyone's personal entitlement to freedom is necessarily limited by another's entitlement to freedom. One of the most fundamental curtailments of personal freedom is imposed by the harm principle, a classical liberal principle, the roots of which go back all the way to John Stuart Mill. The implications of this basic principle are rather strong; as Mill proclaimed: 'whenever, in short, there is a definite damage, or a definite risk of damage, either to an individual or to the public, the case is taken out of the province of liberty and placed in that of morality or law' (Mill, 1859, p. 149). Consequently, while the liberal state should normally refrain from normative intervention in the private lives of individual citizens, a liberal state is morally obliged to intervene, whenever there is a serious (risk of) harm (cf. Wenz, 1988).

In itself the harm principle represents an uncontested component of liberal ethics. Therefore, the principle seems to provide environmental policy and education with a clear and solid ground for justification. An appeal to (the risk of) damage to our natural resources or our common health should be sufficient to justify radical environmental measures. At first glance, such a justification appears to be appropriate as well as easily applicable. Unfortunately, serious problems arise regarding the application of the harm principle. First, the anthropocentric foundation of this principle is problematic, for it precludes a direct state protection of nature. The harm principle protects nature, only if, and insofar as the health or resources of human beings are threatened. Harm to nature itself carries no moral weight whatsoever; whereas our moral intuitions might tell us that it should be the valuing of nature that turns the scale (Vincent 1998).

A second, closely related problem concerns the certainty and provability of harm. The justification of a radical environmental policy and education, directly intervening in the private life choices of individuals, makes great demands on the evidence of (the risk of) harm. With this evidence the state should be able to convince anyone whose private life is somehow influenced by such a policy of the necessity of state intervention. Unfortunately, such convincing evidence is rather exceptional within the context of environmental problems; direct causal links between environmental degradation and, for instance, human health problems are seldom clear and convincing. Those links are predominantly speculative and 'connected through the most tenuous of causal links' (Coglianese, 1998). For instance, it is quite clear that the extinction of animal species will have a disturbing effect on the ecological balances within a particular ecosystem, but the effects on human health or resources are indirect and hard, if not impossible, to prove. Moreover, in the majority of cases, the harm will not be substantial enough to justify state intervention.

Thus, the difficulties of proving harm make it highly problematic to ground strong moral obligations on this principle; the harm is rarely incontrovertibly established and uni-linear, but mostly indirect, insecure, insignificant and probable (Coglianese, 1998, pp. 48–50). However, the seriousness and magnitude of this harm could be disastrous. Therefore, there are strong arguments for a precautionary approach that commits us to protecting any probable harm to the conditions of human survival. Where there are threats of serious or irreversible damage, such an approach would not allow scientific uncertainty to postpone cost-effective measures to prevent environmental degradation. But even such a strong policy would require some proof of probability of harm. The key question we have to face is how much uncertainty or how much risk of harm a liberal state should tolerate its citizens to suffer. Coglianese articulates the question in even more specific terms:

> *Should government take the chance of not responding to what turns out be a real catastrophic risk? Or should it take the chance of interfering with individual life choices for what turns out to have been a false risk of catastrophe? The risk of mistake exists with both action and inaction, and liberal neutrality cannot offer a predetermined solution to such a dilemma (…) Yet liberalism's basic concern with the abuse of individuals by governments, rather than by environmental catastrophes, suggests that it would be more consistent with liberal neutrality to err on the side of less governmental action when such action would impose substantial and unequal constraints on individuals.* (Coglianese, 1998, p. 58)

Coglianese rightly argues that government intervention brings along risks as well as non-intervention does. However, in my view, her final estimation that liberals would probably prefer non-intervention is less self-evident than it may seem. Apparently, Coglianese weighs the risk of individual harm attached to government intervention (abuse of individuals by governments) against the collective risk of environmental harm attached to non-intervention. However, this asymmetry between individual and collective harm is false, since both risks have potential consequences for our personal freedom and autonomy. An environmental disaster might severely limit our personal freedom or even cause a situation in which no individual is capable of leading a free life whatsoever. Therefore, rather than an asymmetric estimation between individual and non-individual harm, the asymmetry is between the immediate and certain harm of government intervention and the long-term and probable harm of non-intervention. This asymmetry does not justify a predetermined commitment to non-intervention, but does make the perceivable preference of liberal governments for the latter option intelligible.

Third, the scope of the harm principle is not only narrow with respect to the range of possible justifications in breadth; it also fails to reach far enough in time. The principle only covers those measures that prevent immediate harm to contemporaries. However, the major environmental concerns of today, such as those concerning the greenhouse effect, the storage of nuclear waste and the preservation of biological diversity, deal with long-term processes of degradation and climate change. The harmful effects of those processes will not reveal themselves within a short period of time, or even within the life span of one or two generations. Moreover, the harm will be done to people living decades, perhaps centuries from now. Therefore, whether the harm principle is of any use within this context depends predominantly on its ability to prevent harm to the life conditions of future generations (Achterberg, 1994b).

Primary goods

Given the inadequacy of the harm principle in this limited form for environmental purposes, liberal theorists have explored the possibilities of the 'positive' idea that underlies the 'negative' harm principle, that is, the idea that there are certain common goods in our society that need to be protected or prevented from harm on behalf of all of us. While liberal theory requires the state to refrain from promoting a particular conception of the good, it does not require the state to remain neutral with respect to all values. There are some values, liberal philosophers argue, which represent the interests of all citizens in general and of democracy in particular. Those values are defined as 'primary goods' (Rawls, 1971, 1993), 'primary values' or 'collective goods' (Raz, 1986). Primary values are distinguished from secondary values, in that they are 'based on benefits and harms that must count as such for all reasonable conceptions of a good life, while secondary values derive from benefits and harms that vary with conceptions of a good life' (Kekes, 1995, p. 19). John Rawls therefore clearly defines primary goods as 'those things that any rational person will want regardless of whatever else he or she wants' (Rawls, 1971, p. 79). For example, democratic 'goods' such as freedom of consciousness, freedom of speech, as well as material 'goods' such as the right to sufficient food, shelter, health service and education should be regarded as primary goods. Contrary to the manifold secondary values, attached to particular conceptions of the good, primary goods are considered to be part of a public morality, actively promoted, distributed and protected by the liberal state. However, in line with the liberal defense of individual freedom against the state, this shared body of primary goods should remain as narrow i.e. minimal as possible. The crucial question here is to what extent environmental 'goods' can be regarded as primary goods, and, as such, be actively promoted and distributed by the liberal state. As suggested before, 'health' and 'natural resources' constitute appropriate candidates for the status of primary good, for both seem preconditional on every life project, whatever its particular circumstances or ultimate goals (Bayles, 1980, pp. 33–34).

While a just distribution of primary goods must be regarded as the nuclear responsibility of the liberal state, the state cannot be held responsible for the distribution of *all* primary goods. For that reason, Rawls makes a distinction between social and natural primary goods. Social primary goods refer to liberties, opportunities, power, income and wealth, the distribution of which is highly correlated to the basic organisation of institutions in society. Natural primary goods, such as health, vigour, intelligence and imagination are not so directly under the control of society. Because natural primary goods cannot be appropriately distributed, Rawls asserts that its distribution is no object of state responsibility. (Rawls, 1999, p. 54–55). This distinction has far-reaching consequences for the inclusion of environmental values in the public liberal morality. For, as summed up earlier, Rawls counts 'health' as one of the natural primary goods. With this rather rough distinction Rawls, intentionally or non-intentionally, assumes that the liberal state has no responsibility whatsoever for the preservation or promotion of public health.

Manning critically responds to this categorisation of health by stressing that Rawls' conclusion seems counter-intuitive to our every day life experiences. It seems implausible to state that public health is entirely beyond the control of society. The national government, for instance, carries immediate responsibility for the public health care and the institutional control over the quality of consumer goods. Furthermore, forms of pollution and environmental degradation strictly controlled and sometimes caused by governmental institutions, can be of direct influence on the health of citizens. Because our health can be positively influenced by public health care, and, conversely, negatively influenced by pollution and degradation, 'health' should be understood, at least partly, as a *social* primary good. Manning convincingly argues that a liberal state is, therefore, morally committed to conduct an environmental policy, aimed at reducing (risks of) harm to human health (Manning, 1981).

Mannings' rescue of the health category from the gloomy status of 'natural primary good', together with the acknowledgement of sufficient natural resources as a social primary good, seems to provide a normative form of environmental education with a solid ground of justification. A liberal state would be allowed to intervene legitimately within the consumption and behaviour patterns of their citizens, whenever those patterns contribute to the depletion of natural resources, the expulsion of greenhouse gasses, the extinction of animal species, the pollution of water, in short, all processes which carry harmful effects on to our health and food resources. When closely examined, the harm principle turns out to form part of the primary goods principle; everyone is entitled to state protection against harm to our health or natural resources. However, the conceptual status of primary goods differs from the status of the harm principle. Whereas the harm principle only leads to a negative obligation (to prevent harm), the primary goods principle determines strong positive obligations, derived from the basic obligation to provide all citizens with the goods 'which any rational person will want'. Thus, the burden of proof is reversed; *citizens* no longer have to prove that their health or resources are threatened, on the contrary; *the state* has to guarantee that the basic needs of citizens are being met.

So, in a way, the harm principle has been incorporated by the primary goods principle. Consequently, most problems discussed within the context of the harm principle, cling to the primary goods principle as well. First, here too, the possibilities of justification are severely restricted by the anthropocentric character of liberal morality. For, presumably, many forms of environmental degradation will hardly have any influence on the means we think necessary for the realisation of our life plan. For instance, the destruction of a particular wildlife area or the extinction of particular animal species will leave our life plans unaffected. Therefore, it is highly questionable whether liberal morality is able to cover the preservation of those natural entities (Musschenga, 1991). Second, the problem of insecurity remains. The primary goods principle requires causal evidence between environmental degradation and human health as well. These problems of provability and risk have been outlined in the previous section.

In fact, we are faced with conflicting primary goods, which have to be balanced against each other. Which primary good should the liberal state give priority? The 'good' of maximum personal freedom (in favour of non-intervention) or the 'good' of a sustainable environment (in favour of intervention)? Unfortunately, experience shows that, generally speaking, long term environmental interests are no match for short term economic interests. Most liberal philosophers therefore recognise that, as long as the basic needs of future generations are not given any moral weight, the primary goods principle will be of little help within the context of the environmental crisis (Singer, 1988, p. 220; Musschenga, 1991).

Obligations to future generations

Both the harm and primary goods principle leave room for an attitude of *après nous le deluge*. As long as we are able to lead meaningful lives and we can be sure the basic needs of contemporaries are being satisfied, then we have lived up to our responsibility. Such an ethic of responsibility would be left empty-handed if people were to decide to foist the harmful effects of our action on to people who will live in the distant future. The analysis of both principles therefore concluded with a hopeful gaze into the future. Is it our responsibility to prevent harm to future generations? Are we obliged to save for posterity, so that their basic needs can be met? Perhaps, we should conceive of ourselves as citizens of an intergenerational community of justice. Consequently, future people should be regarded as no less than fellow citizens, whose 'goods' have to be considered in our present behaviour and choices.

Although there is a widespread agreement about this moral intuition in general terms, its exact implications remain rather unclear. Underlying this persistent vagueness is the lack of any convincing justification of the concept of intergenerational justice, a justification which clearly provides the contents and limits of our obligations to future generations. As John Rawls pointed out in his *Theory of Justice*, the problem of intergenerational justice is extremely difficult to deal with: 'It subjects any ethical theory to severe if not impossible tests' (Rawls, 1999, p. 251). The issue of intergenerational justice is loaded with conceptual intricacies which logically follow from its complex time dimension; the very idea of obligations to future people presupposes the possibility of there being a moral relationship between those living and those not (yet) living. How is it possible to conceive of a relationship between persons who do not live simultaneously? This particular question gives rise to major problems of justification within philosophical ethics. In particular four theoretical problems have to be dealt with: (1) the problem of reciprocity, (2) the problem of ignorance, (3) the problem of paternalism and (4) the non-identity problem. So whether or not a principle of intergenerational justice can be regarded as a necessary clause in the liberal contract ultimately depends on whether these three justificatory problems can be solved.

(1) The problem of reciprocity follows logically from the asymmetrical relationship between contemporary and future generations; our behaviour, our choices will necessarily affect the lives of future generations. They are dependent on the world we leave behind for them. Surely then it is within our power to influence their opportunities, well-being and conditions of life, both positively and negatively. Influences in the opposite direction, however, are logically impossible. We find ourselves in a sovereign position, beyond the reach of future generations. 'Time's arrow' precludes every form of reciprocity. In itself this absence would not be a major problem if the very notion of reciprocity were not considered a defining characteristic of any moral

relationship, at least within this liberal framework. Underlying this characterisation is the idea that the existence of a moral relationship requires a situation in which nobody is invulnerable in such a way, such that he or she does not have to consider the needs and wants of others. Only if the moral parties are mutually dependent on each other, can there be a ground for a moral relationship in the strict sense. Obviously, the intergenerational relationship lacks such a form of mutual dependency (Barry, 1977; de-Shalit, 1995; Rawls, 1999).

(2) That the future cannot be known constitutes another natural feature of time. This feature gives rise to the problem of ignorance. Future generations will always remain 'strangers' to us, as we are ignorant in large measure about their needs and desires. Unfortunately, within philosophical ethics, it tends to be tacitly presupposed that a moral relationship implies a relationship between identifiable individuals or groups. Some philosophers, including such communitarians as Avner de-Shalit and Martin Golding, even argue that the moral parties should be part of a moral community, sharing particular values and ideals (Golding, 1972; de-Shalit, 1995). Without exaggerating this demand of recognisability, as communitarians may seem to do, it should be granted that a lack of all relevant knowledge about the other moral party causes serious problems. For how are we to determine our obligations towards future generations, towards people living centuries from now? How are we to meet their needs and wants when we do not know who they are, what their world will look like and, consequently, what those basic needs and wants will require? Clearly it will be difficult, if not impossible, to determine the content of our obligations to an unknown 'generation X'.

(3) If the content of our obligations to future citizens cannot be based on well-founded knowledge about the needs, wants, values and ideals of future generations, then an intergenerational morality is in danger of becoming paternalistic or even repressive. If we contemporaries determine and pursue a particular future scenario, then we impose our ideals and values upon future generations, and, consequently, deprive them of the opportunity to shape their own world. In line with liberal morality, the basic outline and organisation of the future world should be left to future generations themselves. As is the case with our own contemporaries, they are entitled to pursue their own collective goals and ideals. For this reason, liberal and communitarian critics argue, we should refrain from projecting our values and ideals upon future people (Golding, 1972; de-Shalit, 1995; cf. Rawls, 1999, p. 183).

(4) The so-called *non-identity problem* arises particularly because our present behaviour not merely affects the world and well-being of future generations but their identity as well. Their identity is dependent on our choices and behaviour, here and now. After all, we know that the smallest possible variations in circumstances of conception result in the birth of a completely different person, that is, a person with a different genotype. A different energy-saving policy in the seventies for instance would have resulted in a completely different population in the present. This speculation might appear somewhat far-fetched or even absurd, but, as utilitarians like Parfit indicate, they might have a major impact on the limits of our intergenerational responsibilities. If we understand our responsibility in person-affecting terms – that

is, we regard ourselves responsible for the consequences of our behaviour on the people who are affected by those consequences – then future generations could never hold us responsible for the world we leave behind for them. They are no worse off than they would have been if we were to have acted in their interest, since, if we had acted otherwise then they would not have existed in the first place. Thus Parfit argues that, for those future persons the choice is between existing-under-these-particular-circumstances and not existing at all. Although this objection is mainly directed towards consequentialist theories (the assumptions of which are extremely person-affecting by nature), it will become clear that some contract theories are affected by the non-identity problem as well (Parfit, 1984, pp. 351–379; cf. Grey, 1996; Kavka, 1981).

The inevitable conclusion seems to be that if there is any moral obligation to future generations, then it must be one that urges us to stand totally aloof from all future concerns. 'Hands off!' seems to be the most appropriate maxim. Some philosophers do indeed argue that contemporary generations should refrain from each and every potential interference with the lives of future people, and therefore can have no substantive obligations to them (Golding, 1972; Schwartz, 1978). Nevertheless, the crucial question is whether such an attitude of complete aloofness can logically be maintained. Is it possible to refrain from every influence on future life? The answer to this question must be negative since the world we leave behind will never be a *tabula rasa*. Our policies, choices and behaviour will necessarily affect their conditions of life. Future opportunities will be determined, to a great extent, by the effects of our contemporary actions (or inaction) and choosing (or non-choosing). Therefore, whether we like it or not, whether we think it morally correct or abject, we simply cannot remain innocent. As many authors within this field have stressed, the hands off response is not an adequate one and cannot be maintained. We have to respond to the potential claims that future demands make on us, by virtue of the fact that our behaviour will shape the future world we leave behind after our death. Within a liberal-contractual framework, those demands can only be defended by the immediate stakeholders of future interests: future citizens themselves. Future interests can only be represented in the contract if future citizens are indeed included in the contract community. Consequently, liberal theory cannot simply side-step this issue but has to deal with the problem of including future generations as parties to the contract definition. Thus, what is needed is an adequate answer to the previously outlined problems of justification posed by the issue of intergenerational responsibility.

An extension of Rawls' theory of justice

In order to deal with the previously outlined problems of justification, different authors in the field of environmental ethics and political philosophy take different routes from here. Those who are confined to contractual liberalism, mostly choose to connect their ideas to the influential framework outlined by John Rawls in *A Theory of Justice* (1971)[3] and *Political Liberalism* (1993). From the late seventies, there has been an extensive debate on the implications of his theory for the possibility of extending the liberal contract community, in such a way, as to include future generations. Some liberal philosophers take their bearings from Rawls' *Just Savings Principle*. Others address the issue of intergenerational justice by using formerly unseen possibilities in Rawls' theoretical framework or reformulating its basic assumptions. Before we turn to an exploration of the various attempts in detail, it is necessary to sketch the basic outline of Rawls' theory of justice.

Rawls' main ambition is to define the possible conditions of social cooperation in a 'society of strangers', that is, a society in which citizens do not share a common view on the good life, and where *ipso facto*, any appeal of the government to common values runs the risk of excluding people who do not share this morality. What we need then, according to Rawls, is not a conception of the good, relying on a contested doctrine or morality, but a conception of justice that can count on the hypothetical consent of all citizens, regardless of their conception of the good life. Instead of competing with the comprehensive moral doctrines in society, Rawls wants his principles of justice to precede moral diversity. Moreover, his principles are to make diversity possible. That is, they should define the necessary conditions of humans living together as 'free and equal, rational beings', by guaranteeing the greatest possible freedom of individual citizens to define and pursue their own life-plan, without thwarting the life-plans of others. Rawls articulated this normative premise as the priority of the right over the good. As a qualification of justice 'the right' refers to a situation of 'equal fairness'.

Rawls chooses to establish the priority of the right over the good by deriving his principles of justice from a hypothetical contract, on the content of which all 'rational' beings in his view should be willing to agree. In this contract the basic structure of society and its institutions should be settled. But how can Rawls guarantee that the parties involved are not motivated by private interests that follow from their particular religion, ideology, cultural background, their economic position and similar 'contingencies' that should not determine the content of the contract? In order to safeguard impartiality, Rawls presupposes that the agreement on the principles of justice is reached 'under ideal circumstances'. In order to envisage this presupposition, he presents his famous thought-experiment, in which he defines the initial situation under which the parties are to reach agreement on the principles of justice: the original position. Among other things Rawls suggests that we should envisage the parties in the original position as situated behind a 'veil of ignorance'. With this he assumes that the parties do not know what their characteristics are and under what particular circumstances they live. For instance, they do not know whether they will be man or woman, poor or rich, Catholic or communist, Dutch or Dominican and so on. In short, the parties are deprived of all the knowledge that could possibly enable them to choose principles to their own advantage instead of choosing on the basis of general considerations. Thus, the potential effects of specific contingencies on the definition of justice are nullified.

Furthermore, Rawls assumes that the parties in the original position are 'rational beings', who are motivated by 'moderate selfishness'. This is because he wants his theory to be realistic and therefore derives his principles from weak anthropological assumptions. Since Rawls pretends that he does not presuppose any kind of human altruism, the greatest possible egoist should be willing tot subscribe to the principles defined in the contract. Under the prescribed circumstances, Rawls argues, the parties are likely to reach agreement over two basic principles of justice. The first principle is generally labelled as the equality principle: 'Each person is to have an equal right to the most extensive total system of equal basic liberties compatible with a similar system of liberty for all' (Rawls, 1999, p. 266). The second principle is known as the difference principle: 'Social and economic inequalities are to be arranged so that they are both: (a) to the greatest benefit of the least advantaged, and (b) attached to offices and positions open to all under conditions of fair equality of opportunity' (Rawls, 1971, p. 266).

The issue of intergenerational justice has occupied Rawls from the outset, that is, from the very first draft of his *Theory of Justice* in 1971, when few philosophers had addressed these questions all together. Considerations of justice between generations have contributed to the construction of the theory itself. This will become manifest when we see that core-elements of his theory, such as the design of the original position, his definition of the difference principle and the interpretation of the so-called circumstances of justice, show indications of these attempts to include future generations in his definition of justice. Rawls starts out with an exploration of the human inclination towards time-preference, that is, the inclination to have more regard for some interests rather than others, merely because of ones own location in time. Together with the few predecessors who did address issues of intergenerational justice before, such as the utilitarians Sidgwick and Bentham, Rawls agrees that public judgements which reflect 'pure time preference' are not to be taken into legitimate consideration. In itself, time is not a morally relevant characteristic. However, whereas the utilitarians reject time preference on the basis of its irrationality, Rawls does not admit time preference because this would conflict with the demand of impartiality that is consistent with his understanding of justice as equal fairness (cf. Bayles, 1980, p. 20; Rawls, 1999, pp. 259–260).

Rawls looks at the generation problem from the standpoint of the original position. He requires the parties in the original position to reach agreement about the adoption of a *just savings principle* that insures 'that each generation receives its due from its predecessors and does its fair share for those to come' (Rawls, 1999, p. 254). More precisely, the aim of saving for the future is 'to make possible the conditions needed to establish and preserve a just basic structure over time' (Rawls, 2001, p. 159). In order to reach such an agreement Rawls requires the parties to ask themselves how much 'capital' they would be willing to save at each stage of advance for each succeeding generation, on the assumption that all other generations have saved, or will save, in accordance with the same criterion. In Rawls' view, the savings principle ought to be settled in accordance with the second principle of justice, the difference principle. Among other things, this implies that the social minimum should be settled on a level that maximises the expectations of the least advantaged in each succeeding generation, and without causing hardship to the lowest level of the current society (Rawls, 1999, p. 252). Thus, Rawls argues, the just savings principle should be adopted as a constraint on the difference principle: 'Social and economic inequalities are to be arranged so that they are both: (a) to the greatest benefit of the least advantaged, *consistent with the just savings principle*, and (b) attached to offices and positions open to all under conditions of fair equality of opportunity' (Rawls, 1971, p. 266). Positively stated, the just savings principle can be formulated as an obligation to 'save for each succeeding generation without causing hardship to the lowest level of the current society' (Manning, 1981, p. 292).

Whereas the just savings principle is easily settled in sympathy with our intuitive sense of justice, Rawls encounters immense complexities in his justification of this principle within the delicate theoretical web he has spun. At the very moment he has dealt with one contradiction in his theory, he comes into conflict with another basic assumption or intuition. However, Rawls and his critics go to great length, considering the implications of the particular design of the original position and reconsidering its basic assumptions in the light of a just savings principle. In order to secure an impartial standpoint that does not leave room for time preference, Rawls assumes that the parties in the original position are not informed about the particular generation they belong to. But since the application of the principle of justice is dependent on a condition of mutual reciprocity and dependency of the contractants, he chooses for a 'present-time-of-entry-interpretation' of the original position. That means that, although knowledge about the generation one belongs to is located behind the veil of ignorance, the parties do know that they are contemporaries. Complete ignorance about one's own generation and the contracting parties one negotiates with, would not only do injustice to the necessary condition of mutual dependency, but would, in Rawls' view, ask too much of our imagination as well. What does this imply for the motivation of the contracting parties? Unfortunately, as Rawls is ready to admit, there is nothing to prevent the contemporaries behind the veil of ignorance from choosing their particular interest now. Nothing prevents them from using up the natural resources, and consuming all the available savings. Nothing prevents them from time preference because, as contemporaries-among-contemporaries, they remain immune to the claims of other generations. There is no mutual reciprocity that forces the contracting parties to consider one another's interests. '(…) so unless we modify our initial assumptions, there is no reason for them to agree to any saving whatever. Earlier generations will have either saved or not; there is nothing the parties can do to affect that' (Rawls, 1999, p. 254–255).

Rawls seeks to overcome the deadlock in the negotiations between the contracting parties by adopting a *motivational assumption*. That is, he assumes that all the parties represent family lines, who care for the well-being of their descendants, at least for the more immediate ones, their children and grandchildren. This sentiment could be thought of as parental concern: 'We can adopt a motivation assumption and think of the parties as representing a continuing line of claims. For example, we can assume that they are heads of families and therefore have a desire to further the well-being of at least their more immediate descendants' (Rawls, 1999, p. 111). To this motivational assumption he adds another assumption 'that the principle adopted must be such that they wish all earlier generations to have followed it' (Rawls, 1999, p. 254). These two assumptions, together with the operation of the veil of ignorance, Rawls argues, will guarantee that 'any one generation looks out for all' (Rawls, 1999, p. 254). Conclusively, the savings principle not only covers an emotional span of two or three generations, but, this commitment is also universalised by the additional assumption and the veil of ignorance into a 'chain of commitment' that secures the adoption of a transgenerational savings principle in the original position. By deliberating about a just amount of savings, citizens are obliged to represent the interests of the future members of their family line, not only of themselves.

The circumstance of justice

Whereas Rawls himself is convinced that his justification of a principle of intergenerational justice is established his interpreters remain sceptical and divided. In the first place, there is thorough debate on the question whether an empirical psychological postulate, like the motivational assumption, can ever count as a legitimate argument in the justification of an ethical or political principle. On this particular point, Brian Barry compares Rawls with a magician 'putting a rabbit in a hat, taking it out again and expecting a round of applause (…) if it is acceptable to introduce desires for the welfare of immediate descendants into the original position simply in order to get them out again as obligations, what grounds can there be for refusing to put into the original position a desire for the welfare of at least some contemporaries?' (Barry, 1977, p. 279–280). Barry sharply points at the fact that the content of the motivational assumption seems at odds with the characterisation of the parties in the original position as 'non-altruistic', 'mutual disinterested'. The strong and wide appeal of Rawls' theory is mainly due to the fact that he derives strong obligations from weak assumptions concerning human nature. In fact, 'moderate selfishness' is one of the so-called 'circumstances of justice' that Rawls borrows from David Hume in order to describe the 'normal conditions' under which human cooperation is both possible and necessary, and thus, a conception of justice is needed. When these circumstances do not hold, or one of them does not hold, the principle of justice can find no application. Rawls is ready to admit that the motivational assumption comprises a contested element in his theory, but he nevertheless chooses to hold on to this assumption because, in his view, it requires no more benevolence of the parties than one can plausibly assume to be part of human nature. Among other critics, Manning strongly opposes this suggestion: 'A good case can be made for the idea that parental care is not a part of natural human psychology but is a cultural phenomenon. Parents have been known to sacrifice their children to gods and sell them into slavery. Child abuse and neglect is only just short of being rampant in our own society. Parental love does not seem to be a strong enough motive on which to base the savings principle. Again it appears best to stay with mutual disinterest' (Manning, 1981, p. 163; Barry, 1977; cf. Thero, 1995).

In the second place, critics point at the lack of reciprocity in the relationship between generations. According to the 'circumstances of justice' mentioned earlier, a particular kind of reciprocity between the contracting parties is required, if the principle of justice is to be applied. More precisely, it is the condition of 'relative equality' that requires the parties to be 'roughly similar in physical and mental powers; or at any rate, their capacities are comparable in that no one among them can dominate the rest. They are vulnerable to attack, and all are subject to having their plans blocked by the united force of others' (Rawls, 1999, p. 110). Similar to other contractual theorists, Rawls tries to explain the 'willingness to cooperate' by assuming a condition of mutual dependency. Underlying this postulation of a reciprocal relationship is the idea that the survival of a strong moral contract requires a situation in which nobody is invulnerable in such a way they do *not* have to consider the needs and wants of other(s). Only if the moral parties are mutually dependent on each other, can there be the grounds for a moral relationship in the strict sense. This particular requirement reflects the non-moral worldview expressed by Hobbes in his *Leviathan*. Justice is necessary because, and only where, there is a situation in which everyone is a threat to everyone else. Here, the motivation to sign the contract is based on mutual fear, and justice is understood as penal machinery which serves as men's security against one another (Hobbes, 1960; Barry, 1978; Rawls, 1999).

Although the case of Hobbes is extreme, even Rawls suggests that social cooperation and justice are impossible without a certain degree of reciprocity. His design of the moral contract is based on the psychological assumption that people are likely to act justly only regarding those who can give justice in return (Rawls, 1999, p. 447). Obviously, the intergenerational relationship lacks any form of such mutual dependency and reciprocity. Therefore, to put the conclusion as simply as possible: because future generations cannot consider our needs, we are not obliged to consider theirs. Rawls, however, maintains that there is a particular kind of reciprocity in our relationship with future and past generations that should be conceived of in terms of 'equal compensation'; each generation finds itself located in a chain of giving and receiving. Ideally, a fair balance is achieved between the inheritance we receive from our predecessors, and the inheritance we leave behind for our grandchildren: 'The only economic exchanges between generations are, so to speak, virtual ones, that is, compensating adjustments that can be made in the original position when a just savings principle is adopted' (Rawls, 1999). Most critics, however, are not convinced by this virtual solution since this particular kind of 'reciprocity' can only serve its justificatory purpose in a form that would stretch our imagination too far.

As we have seen, Rawls comes into conflict with the requirements of his 'circumstances of justice' twice. First, the motivational assumption conflicts with the circumstance of 'moderate selfishness', and second, the very idea of a just relationship between generations conflicts with the circumstance of 'relative equality of power'. Beyond these two circumstances, there is the third and last circumstance of 'moderate scarcity'. Whether this circumstance will hold is not clear either, because it is not inconceivable that in the future, particular resources will be *severely* scarce. The provisional conclusion that can be drawn from this brief analysis is that, since these circumstances do not obtain within the intergenerational context, the principle of justice can find no application. For those liberal philosophers, who were not ready to accept this conclusion, this was an incentive to examine the status and implications of these 'circumstances of justice' more closely.

Rawls borrowed the 'circumstances of justice' from David Hume in order to describe the background conditions that give rise to the necessity of reaching an agreement over a principle of justice. More precisely, they are to describe the 'normal conditions' under which human cooperation is both possible and necessary, and thus, a conception of justice is needed. The 'circumstances' are to ensure that the theory of justice rests on weak and realistic premises. It is his aim to incorporate widely shared and yet weak conditions by deriving all duties and obligations from minimal assumptions concerning moral behaviour. A conception of justice should not presuppose extensive ties of moral sentiment or self-sacrifice. At the basis of his theory Rawls tries to assume as little as possible. Not a spark of altruism is expected from the contracting parties. Their choices and behaviour are supposed to be guided by 'mutual disinterestedness'. The main idea is that the application of a principle of justice requires a society that is marked by a conflict as well as an identity of interests. Rawls for instance, embraces the condition of moderate scarcity because, in his view, a principle of justice can find no grip in extreme situations: in a situation of superabundance, there is no conflict of interests, and therefore no need to agree on a principle of distribution since there is enough for everybody. Justice would be a redundant and idle ceremony. In the opposite situation of extreme scarcity, on the contrary, there is no identity of interests, since people will be motivated by a struggle for survival. For this reason, Rawls and Hume argue, justice only applies under 'normal conditions of 'moderate scarcity'. Together with the circumstances of relative equality and moderate selfishness, one can say, in brief, 'that the circumstances of justice obtain whenever persons put forward conflicting claims to the division of social advantages under conditions of moderate scarcity' (Rawls, 1999, p. 110).

Barry argues that the circumstances of justice comprise a hybrid element in Rawls' theory. He illustrates his argument by pointing at the absurd implications of the requirement of relative equality. In a society, where the political system denies its citizens equal rights and access to democratic institutions, like the former apartheid regime in South Africa, the theory of justice would not be able to offer human rights activists an independent argument in favour of justice. Rawls and Hume would have to leave them empty-handed, because in their society, the circumstance of relative equality does not obtain, and *ipso facto*, the principle of justice cannot be employed to criticise racial politics. This conclusion is absurd, thus Barry argues. The circumstances of justice lead to a 'hollow mockery of the idea of justice – adding insult to injury. Justice is normally thought of not ceasing to be relevant in conditions of extreme inequality in power but, rather, as being especially relevant in such conditions' (Barry, 1978, p. 222). Similarly, environmental problems usually emerge in situations where natural resources are scarce. If a principle of justice can find no application in situations of extreme scarcity, a theory of justice would be of little help. In line with this criticism, Van der Wal argues that Rawls employs an awkward theoretical construction by implying the conditions of realisation within the definition of justice itself. Together with Barry, Van der Wal states that, whereas in practice particular societal factors have proven to be of decisive importance in the emancipation of socially deprived groups and the establishment of basic right, the legitimacy and content of a principle of justice should never be made dependent on its empirical presence in a given society (Van der Wal, 1979, p. 21).

Previous considerations have lead Barry to redefine the status of the circumstances of justice. He agrees with Rawls that an ideal of justice can only arise in a society where people are more or less equal, and that it is likely to presume that justice will be established in situations of relative equality, moderate selfishness and moderate scarcity. But contra Rawls, Barry rightly argues that this does not imply that the legitimacy of this idea of justice is limited to these circumstances of justice, and that the idea of justice cannot be applied to a situation in which the circumstances do not yet hold. If this were so, then the theory would imply an *apriori* justification of the present status quo. Therefore, Barry suggests that we should not understand the circumstances of justice as *necessary* conditions, as Rawls proposes, but as *sufficient* conditions for the application of a principle of justice. The bare fact that the conditions of moderate selfishness, relative equality of powers and moderate scarcity obtain, is in itself a sufficient reason to reach agreement over a common conception of justice (Barry, 1978, p. 225). Moreover, it is likely to assume that such an agreement is easier to establish in this situation than in a situation where these circumstances do not hold, but the principles of justice in terms of equal fairness do not lose their validity in 'extreme situations'. On the contrary, in circumstances of extreme selfishness, inequality of powers between the parties and extreme scarcity, the principles of justice will have to prove what they are worth, even though justice will indeed be harder to establish. Thus, Barry defends the principles of justice for better *and* for worse.

In order to understand more deeply why this modification is necessary, Barry goes on by stating that Rawls' misconception of the circumstances of justice is due to an untenable mixture of Humean and Kantian elements in his theory. On the one hand, Rawls chooses to hold on to the Kantian idea of hypothetical universalisability in his design of the original position, while, on the other hand, he introduces a conventional conception of justice by introducing the Humean circumstances of justice in his theory. Hume defined these circumstances in his effort to demonstrate that the virtue of justice is a pure artefact. In his view, rules of justice are conventional. They are determined by the particular circumstances in which they arise, and in his view, there is no external standard of justice against which the conventional rules can be assessed. For this reason, a law cannot be unjust, since it is the law itself that defines justice. A conception of justice is no more than a function of a particular kind of human cooperation that should lead to mutual advantage, thus Hume argues. Needless to say that this *conventional* understanding of justice is at odds with the *universal* and transcendental Kantian understanding of justice. Thus, the Humean circumstances of justice are at odds with the universal pretensions underlying the idea of hypothetical consent in the experiment of the original position. Therefore, Barry argues, it is not only on behalf of future generations but for reasons of consistency as well, that the status of the circumstances of justice needs to be amended. With the proposed understanding of the circumstance of justice as sufficient conditions, Barry wants to do justice to the Humean intuition that rules of justice are conventions (in contrast to the principles of justice) that particularly arise in situations where there is a conflict as well as an identity of interests. On the hand, however, he claims not to undermine the universal claim of validity underlying Rawls' principles of justice (Barry, 1978; Achterberg, 1994b). Whereas one might interpret this solution as a fair balance between Humean and Kantian elements, in my view, the Humean intuition is being disarmed and put under guidance of the Kantian law. After all, the underlying Humean point that any principle of justice should be conceived of as a conventional artefact becomes insignificant or even self-refuting as soon as it is subsumed under the predicate of universalisability.

A negative principle of intergenerational justice

Barry's adjustment of Rawls' theory of justice has been adopted by many liberal philosophers in the field of environmental ethics[4]. These philosophers argue that, as a result of this adjustment, the circumstances of justice do not obstruct the assignment of rights to future generations anymore. Among them there is, however, controversy concerning the question what the exact nature of their rights should be, and consequently, how we should include future generations in our hypothetical dialogue on environmental issues. The question is whether the savings principle – as a constraint on the difference principle – is the most adequate way of framing our obligations to future generations. The savings principle requires us to relate to future generations as hypothetical fellow citizens, who might belong to the least advantaged of our society. Therefore, they have a legitimate claim on our savings, and we have an obligation to make the deprived ones among them better off. In a way, the savings principle (as a constraint on the difference principle) postulates a particular kind of solidarity between generations, in which the protection of the weak and vulnerable is central (cf. Achterberg, 1994).

Barry as well as Hilhorst remark that Rawls deals with the issue of intergenerational justice from a somewhat one-sided angle of distributive justice. The environmental crisis cannot be fully understood as a problem of unequal resource distribution, but includes other problems of depletion, problems of degradation and pollution as well. The environmental problems that hang together with the greenhouse effect, for instance or the problems of acidification or the storage of nuclear waste can hardly be conceived of as problems of distribution. Whereas the label of 'unequal resource distribution' does not cover all environmental problems, they can altogether be regarded as problems that might be harmful to the basic rights of future citizens. Therefore, one might argue that environmental problems require a negative duty to prevent harm to the fundamental life-conditions of future generations. Barry arrives at a similar conclusion: '(…) if we concentrate on the question how much we are obliged to make our successors better off, we miss the whole question whether there may not be an obligation to avoid harm that is stronger than any obligation to make better off' (Barry, 1977, p. 267).

In contrast to a positive obligation to make the least advantaged of our successors better off, which aims at establishing a particular favoured condition in the future, this negative morality aims at minimising our interference with the lives of future generations, or rather, put in positive terms, maximising their range of opportunities. Such a morality could be guided by the following maxim: do not bring about any state of affairs that is irreversible or irrevocable, that restricts the range of options and opportunities of future generations. In short, leave the future open to future generations. This maxim requires that present generations do not interfere with the content of future choices, but rather, that they bring about the necessary conditions of future freedom of opportunity[5]. Consequently, it is our obligation to prevent the exclusion of prevailing options and opportunities in the future (Barry, 1977; Bayles, 1980; Routley, 1982; cf. Achterberg, 1994). In fact, this is exactly what the concept of sustainable development is about: meeting 'the needs of the present without compromising the ability of future generations to meet their own needs' (WCED, 1987, p. 12).

Many liberal philosophers in the field of environmental education therefore drop the difference principle as a 'hang-up' for our obligations to future generations in terms of intergenerational solidarity. Some of them choose to follow the first principle of justice instead: the equality principle, that requires each person 'to have an equal right to the most extensive total system of equal basic liberties compatible with a similar system of liberty for all' (Rawls, 1999, p. 266). A negative principle of intergenerational justice that takes its bearings from the equality principle might be defined as Achterberg does, that is, in terms of *a fair equality of opportunities*: 'in their opportunities and possibilities of leading their lives in accordance with their own conception of the good, in their opportunities of using natural resources, and in their opportunities of enjoying nature, later generations should not be "worse off" than we are' (freely translated from: Achterberg, 1994). In similar terms the utilitarian Bayles defines an *equivalent principle*, that requires us to compare and balance their quality of life against our quality of life: 'The present generation has a duty not to render it substantially unlikely that future generations can have an indefinitely sustainable quality of life as high as it has, an equivalent quality of life' (Bayles, 1980, p. 21). Besides, the application of the equality principle leaves more room for the prevention of harm, as Richard and Val Routley illustrate with their *transfer limiting principle*: 'one is not, in general, entitled to simply transfer the costs of a significant kind arising from an activity which benefits oneself onto other parties who are not involved in the activity and who are not beneficiaries' (Routley, 1978, p. 123). From this principle they derive the so-called transmission *principle*: 'we should not hand the world we have so exploited on to our successors in substantially worse shape than we "received" it'. After all, that would imply a significant transfer of costs (Routley, 1978, p. 123; cf. Achterberg, 1989, p. 103). Despite, the different designs of a principle of intergenerational justice, the definitions just referred to all have in common that they aim to establish a continuing 'chain of obligations' between succeeding generations (cf. Howarth, 1992).

If intergenerational justice can be understood in this 'negative' way, at least some of the major objections seem to be superseded. First, the objection of paternalism does not hold, because this morality is evidently *negative*. Contemporaries are not supposed to fill in the future for future generations, but, nevertheless, they do provide the necessary conditions that will enable those generations to give shape to their own lives. By keeping the future open, we make future life possible, without imposing our own values and ideals onto people in the future. Second, by adopting a negative morality, the problem of ignorance seems to be superseded as well. For this, morality requires no particular knowledge about the values, needs and ideals of future generations, or it does so only in a minimal sense. Moreover, the only thing we assume, is that they will need personal freedom and such goods as drinking water, clean air and natural resources, sufficient to satisfy their basic needs. As outlined before, these goods are regarded as 'those things that any rational person will want regardless of whatever else he or she wants' (Rawls, 1999, p. 79). At this point Brian Barry somewhat cynically retorts to any protestations of ignorance the objection of ignorance: 'Of course, we don't know what the precise tastes of our remote descendants will be, but they are unlikely to include a desire for skin cancer, soil erosion or the inundation of all low-lying areas as the result of the melting of the ice-caps. And, other things being equal, the interests of future generations cannot be harmed by our leaving them more choices rather than fewer' (Barry, 1977, p. 274).

Whereas liberal philosophers have successfully dealt with the problems of ignorance and paternalism within their particular (Rawlsian) framework, the *non-identity problem* requires a separate answer. Because those liberal philosophers in the field of environmental ethics who follow the approach of a negative principle of intergenerational justice hold on to Rawls' construction of the original position and the connected idea of hypothetical consent, the *non-identity problem* takes a specific form. How can we choose policies at one and the same time that will affect the identity of the people who will live, and ask the not-yet-born to join us in taking these decisions?[6] This is only possible by allowing *potential* persons to enter the negotiations behind the veil of ignorance, that is, people who might or might not exist, dependent on the policies the parties decide upon. So, the parties have to decide whether the potential persons among them should be brought into existence or remain 'potential'. Who are the people we are talking with, here, and who are we talking about? Due to the contingent status of future generations, the parties in the original position have to deal with more absurd problems such as these. Obviously, the thought experiment of the original position, that seemed to be very clear and helpful in determining our responsibilities, requires too much of our abstraction and imagination as soon as those complexities are involved in the negotiations behind the veil of ignorance (de-Shalit, 1995).

But is it necessary to introduce the contingency of existence into the original position? In other words, do Rawls and his fellows assume a person affecting responsibility? Many liberal philosophers have been puzzled about this question. Howarth, for instance, suggests that we should think of our obligations to future generations as mediated by a chain of obligations. The obligations to my great-grandchildren with whom I do not live together but whose identity is contingent upon my behaviour are, for instance, mediated by my obligations to my children, my children's' obligations to my grandchildren and my grandchildren's' obligations to my great-grandchildren. Thus, a chain of causal obligations is established, spanning an infinite number of generations, thereby expanding our person-affecting responsibility into the indeterminate future (Howarth, 1992). Whereas Howarth succeeds in side-stepping the non-identity problem, another problem takes its place. Does a multi-levelled approach such as this, offer sufficient ground for our obligations to those who will be a hundred generations away from us? A certain 'discount' of our obligations into the remote future seems inevitable; the more 'chains' in between us and future people, the less weighty our responsibilities will be. In the end, our responsibilities will be nullified or outweighed other responsibilities.

The most promising attempt to deal with the non-identity problem is, in my view, offered by Hilhorst, who argues that our obligations to future generations need to not be grounded in person-affecting relationships. That is to say, as long as we see our intergenerational responsibilities in terms of obligations to prevent harm, then we do have to prove how future people are affected by the consequences of our present behaviour, since the concept of harm is defined by one agent affecting another agent. In principle we cannot hold someone responsible for our harm who did not personally harm us. Contrary to the concept of harm, the concept of 'wrong' does not imply a relationship that is mediated by personal consequences. On the contrary, were someone to leave behind a time-bomb in her car, her action should be regarded 'wrong', irrespective of who will suffer from the explosion. The same might be argued in case of the storage of nuclear waste. We do not need to listen to the complaint of identifiable future complainants in order to condemn our current policy on nuclear energy at this point as 'wrong'. Hilhorst argues that we should judge in similar terms those unsustainable policies that will confront some future individuals – whoever they might be – with a *fait accompli* of the kind Parfit sketched before; either exist-under-these-horrible-circumstances or do not exist at all. To put someone in a situation like this is conceptually tantamount to blackmail. One forces someone into a position in which he has no other choice than to accept the current situation to which he strongly objects. In the construction of this argument Hilhorst employed MacLeans' concept of the *place-holder complainant*. In our present choices we have to anticipate the potential compliant of the *place-holder complainants*, that is those people, whoever they may turn out to be, who are not being personally harmed, but who rightfully complain about their predicament, namely, the horrible position they were forced to take. By using this argument, Hilhorst opts for an alternative, non-personal way of dealing with our responsibility to future generations, in such a way, that the non-identity problem does not affect the justification anymore (Hilhorst 1987, pp. 76–82).

Reciprocity and motivation

Whereas the problems of ignorance, paternalism and the non-identity problem seem to be met by this negative morality approach, the problem of reciprocity remains unsolved; there simply cannot be a reciprocal exchange of 'goods' and 'bads'[7] between contemporary and future generations. The intergenerational relationship allows for exchange in one way only. Whether we do good or bad to future generations, we will neither receive rewards nor suffer repercussions for our present behaviour. Our moral choices are left unsanctioned. The crucial question is whether the absence of reciprocity precludes the acknowledgement of a moral relationship between generations within a liberal contractarian framework. To answer this question, it is necessary to take a close look at the liberal-contractual understanding of moral relationships again, but now in the light of the suggested reconceptualisation of the circumstances of justice. Barry and like-minded philosophers who adopted this negative morality approach claim that nothing obstructs the adoption of an intergenerational principle of justice, now that the circumstantial requirements of moderate scarcity, relative equality and moderate selfishness are reconceptualised as sufficient, rather than necessary conditions. Theoretically speaking, this reconceptualisation comprises an ingenious exercise in order to side-step the identified problems of justification. But if we examine what the practical implications are, it will become clear that this modification affects the very ambition of the theory of justice.

If we concentrate on the circumstance of 'relative equality,' for instance, what does it mean that the requirement of reciprocity is not considered a necessary but a sufficient condition? In my view, this modification implies the acknowledgement that we are likely to draw our moral motivation from reciprocal relationships, since reciprocal relationships provide us with a strong self-interest to do justice to the other, but we are nevertheless obliged to follow the principle of justice in non-reciprocal relationships as well. That is, we are expected to apply our principle of justice in situations where we are not motivated by self-interest. At this point, the implications of Barry's modification conflict with the 'weak assumptions' which Rawls aims to secure through his design of the original position. Behind the veil of ignorance, the parties are not expected to act out of altruism or benevolence. In sharp contrast to this, such an altruistic motivation is precisely what is required of citizens after the veil is lifted. Citizens are expected to draw their motivation from the intergenerational solidarity they experience within their present community of interdependent citizens, and apply this motivation to non-reciprocal relationships like those with future people. Obviously, this is not what Rawls had in mind when he designed his theory. For him, this would be a clear-cut example of overcharging human benevolence. So, by modifying the circumstances of justice, Barry undermines the weak assumptions and *ipso facto* the wide appeal of Rawls' theory of justice. In fact, one might argue that Barry obscures the very ground that enables Rawls to expect compliance of every citizen with his principles of justice. My criticism on this point reflects Sandel's criticism on the discrepancy in Rawls' theory between the weak assumptions concerning human nature in theory and the strong solidarity required by its application in practice (Sandel, 1982).

Slightly different from what Barry suggests, Hume formulated 'the circumstances' to limit the sphere of justice in favour of the social sphere of solidarity and mutual benevolence. For Hume, justice is a relevant virtue, only in extreme situations, marked by scarcity and the absence of affective ties. In most situations however, where these circumstances do not obtain, we better trust on our moral sentiment and 'natural' motivation. Hume argues that too strong an emphasis on distributive justice will ruin ties of solidarity (Taylor, 1995, p. 20). Rawls, however, employs the circumstances of justice to assure his 'weak assumptions', but simultaneously aims at extending the sphere of justice, and *ipso facto* the sphere of mutual disinterestedness. As the foregoing indicates, Rawls' theory of justice cannot do without the circumstances of justice, conceived of as necessary conditions for its application. Whereas the striving for inclusion of future generations urges us to drop the circumstances of justice, the application of the theory requires us to reintroduce them again. Thus, any liberal-contractual justification of obligations towards future generations as hypothetical fellow citizens will break down on the problem of motivation. An intergenerational principle of justice, like that of sustainable development, cannot be justified on liberal grounds. The relationship of justice holds only among contemporaries. Respecting future generations is not a matter of justice *between* generations. If there is any place for a principle of justice here, it is a matter of justice *with respect* to the future.

Even if the lack of motivation would not preclude the justification of an intergenerational principle of justice, this would comprise a practical problem of its own. What inspires individual citizens to take responsibility for the well-being of future generations? How can they be motivated to contribute to the establishment of a 'sustainable world'? The underlying motivation would have to be quite strong, for our contribution to a sustainable future might require serious sacrifices. Liberal political theory has not yet given an adequate answer to this problem of personal commitment. Why should people commit themselves to an abstract principle of intergenerational justice? Rather than by their commitment to procedural-democratic values and responsibilities, citizens will be motivated by their commitment to the traditions, ideals and projects of the community they are part of and to which they feel intrinsically bound. Such a personal commitment to a conception of the good life probably provides a stronger motivation to contribute to environmental ideals than a rather abstract idea of an intergenerational community of justice. These communal sources of morality have a central place in communitarian theory. Therefore, in the next part of this chapter the communitarian view of intergenerational responsibility will be examined (cf. de-Shalit, 1995).

The metaphor of the free market

The previous problems of justification and motivation cannot easily be done away with; they are not incidental or merely contingent. The assumptions of reciprocity and mutual disinterestedness are constituent parts of the liberal way of thinking about public morality and politics. It is obvious then that the liberal framework itself constitutes the problem here. Its language and assumptions are not appropriate to deal with these kinds of problems, and this in turn raises more general questions about the liberal framework itself. Moreover, the assumptions of reciprocity and mutual disinterestedness clearly point towards the metaphor of the free market, inherent in every contract theory. Although they vary in purpose and outline, contract theories share at least one basic assumption as far as the nature of morality is concerned: morality is grounded in a rational consensus established in a bargaining process involving all moral parties. The manifold contract theories differ with respect to the question of the *object* of consensus. Economic liberals such as David Gauthier argue that bargaining is concerned with the mutual protection of 'rational egoists' (cf. Gauthier, 1963). Here, the free market metaphor is clear. Though Rawls' contract theory concentrates less narrowly on economic interests, and refers to a just distribution of social rights and obligations, his principles of justice are still the results of a contractual exchange of 'goods' and 'bads'. Within these contractual terms, moral and political issues are regarded as problems of *distributive* justice. Even the environmental crisis is conceived of as a problem of property distribution, or more precisely, as a problem of *resource*- and *risk management*. Furthermore, interactions between citizens and between citizen and state represent the structure of the trade relation. Interactions are determined by 'calculation', 'exchange' and 'bargain'. In fact, the very existence of the contract is the result of such a bargain: citizens pay the 'price' of their freedom for the 'good' of a well-ordered society. In exchange for the restriction on their freedom they receive the legal status of citizen.

The metaphor of the free market gives rise to a rather narrow understanding of rational choice and behaviour, as represented behind the 'veil of ignorance'. Rawls thinks of the parties in the original position as 'rational in the narrow sense' and mutually disinterested: '(…) they are conceived as not taking an interest in one another's interests (…). Moreover, the concept of rationality must be interpreted as far as possible in the narrow sense, standard in economic theory, of taking the most effective means to given ends' (Rawls, 1999, p. 12). Rawls envisages the political community as a collection of discernible sovereign subjects, standing apart, merely involved in realising their own life plans. As such, the original position postulates a particular notion of the moral person, conceived as prior to society. By means of this thought experiment individual persons are detached from the very sources – social and power relations, their natural environment, language community, culture and the whole set of practices – that constitute their 'selves' and are positioned towards these 'contingencies' in an instrumental way, as if we stand in relation to those things as 'unencumbered choosers' looking for the best option. Liberal theory in general aims to safeguard our personal autonomy by urging us to control all 'external influences and claims' that could possibly affect our ways of acting and thinking in a way we might not approve of. Paradoxically, by doing this, we run the risk of losing our receptivity to the appeal things make on us and cutting ourselves off from the authentic sources of meaning – that are located beyond the 'self'. By postulating that 'the self is prior to his ends' Rawls creates a distance between me and the constituent parts of my identification in a way that ignores the existential nature of morality and judgement. This line of criticism will be examined more closely in the next section (cf. Oldfield, 1990, p. 19–20; Taylor, 1992; Smith, 1998; Rawls, 1999, p. 492).

The Rawlsian picture of the moral person clearly reflects the solipsistic *cogito* of Descartes and the rational Kantian subject, whose glorification of the subject's autonomy has been widely challenged. Hannah Arendt argued, for instance, that every personal act or judgement is enacted within an already existing 'web of relationships'. This intersubjective level of meaning precedes every subjective rationality: 'Although everybody started his life by inserting himself into the human world through action and speech, nobody is the author or producer of his own life story. In other words, the stories, the results of action and speech, reveal an agent, but this agent is not an author or producer. Somebody began it and is its subject in the twofold sense of the word, namely, its actor and sufferer, but nobody is its author' (Arendt, 1958b, p. 184). If our individual life stories and the web of human relationships are mutually interwoven, as Arendt suggests, then, obviously, there is more to society than a loose collection of individual choosers, and there is more to the common good than the mere addition of individual ends, tastes and preferences.

Another source of concern arises over the fact that every market economy is in need of a strict division between public and private matters. The basic economic activities, those of production, bargain and trade, can only serve their public purposes if there is a private sphere where people are to consume, buy and reproduce (cf. Arendt, 1958b, pp. 79–135). Analogously, Rawls' hypothetical contract between 'mutually disinterested' persons draws a sharp distinction between the status of private and public concerns, that is, between contractual and non-contractual concerns. In general, the liberal contract is designed to keep the public requirements of citizenship as minimal as possible in order to maximise everyone's personal freedom within the private sphere. Thus, a priority is established; private concerns are privileged over political concerns. Moreover, political concerns are regarded as a function of private concerns. Political debate and judgement are denied intrinsic value because they are to be evaluated in terms of their contribution to the private pursuit of happiness. By setting apart the public and private like this, individuals and groups in society are posited towards the common good and common concerns in an instrumental way. Political action and involvement are presented as instruments to achieve private goods. The entitlements we secure on a public level are to be 'consumed' within the private sphere. This is not to say that an intrinsic commitment to the public good is impossible. Of course, a public spirit and commitment might emerge in this process, but it is not required nor rewarded by the political system[8]. In this sense, the environmental crisis is only politically meaningful insofar as it threatens the freedom we experience in the private sphere. Public requirements appeal to our responsibility as consumers rather than citizens.

The liberal state is said to aim at enabling individuals to develop, choose and pursue their own conception of the good or lifeplan, without interfering in these moral disputes themselves. Above all, this presumption raises questions concerning the possibility of such a non-ideological stance. Is it possible to ground a common allegiance to a contract on the preferences and needs of a pre-political subject (like Rawls' parties in the original position) before they have entered public life? Is it possible to anticipate ones needs apart from one's place in society, one's ideology or religion, one's family situation, lifestyle – in short – all things that could possibly render meaning to those needs? And consequently, does a liberal morality not falsely claim a second-order status? Though very interesting, I do not want to pursue this question here any further because they will be elaborated upon in section 2.3. Here, I would like to put forward more practical objections to this sharp distinction between the private and public spheres, between individual and citizen, and between first-order and second-order morality.

My first objections reflect the communitarian criticism that liberal philosophy cannot cope with the major societal problem we are confronted with in late- or postmodern society. As the previous analysis shows, a majority of environmental concerns can hardly be dealt with in the public arena because moral concerns for future generations are regarded to be of a private and first-order rather than public in nature. It is interesting to note that in almost every religion, world-view, ideology or lifestyle there are points of application for an intergenerational responsibility. Christians, for instance, find their environmental responsibility grounded in the biblical imperative of 'stewardship' or the Franciscan respect for animal life, socialists draw from Marx's ideas about the marginalisation of nature, and other people will start from a vegetarian or bio-dynamic lifestyle and so on. However, moral commitments and arguments of these first-order kinds should not, according to liberal theory, play a part in the public debate. The same goes for arguments concerning the intrinsic value of nature in debates on biotechnology and genetic engineering, or arguments concerning the 'rights' and welfare of animals; although a majority of people in modern society will acknowledge that animals are entitled to a certain moral consideration and think of the value of nature in terms beyond those of human resources. These considerations are generally denied status within the contract and excluded from political debate because they are regarded as 'metaphysical'[9]. In short, the moral concerns people find themselves intrinsically bound to are taken off the public agenda and left to private device. Thus, liberal politics give rise to a privatisation of moral dilemmas.

This narrowing of the public agenda might seem to be at odds with the main 'spirit' of liberalism, which is said to glorify an open exchange of ideas and opinions in society. However, in practice, liberal theories do tend to restrict room for public debate by using contractual principles as 'principles of preclusion'. According to the analysis of Gutman and Thompson these principles are to 'preclude fundamental moral conflict by denying certain reasons moral standing in the policy-making process' (Gutman and Thompson, 1990, p. 125). They serve their purpose by determining which issues deserve a place on the political agenda in the sense of being a legitimate subject for legislation. The aim of precluding conflict might narrow the agenda in such a way, that it leads to an impoverishment of public debate. Therefore Gutman and Thompson argue in favour of a 'public philosophy of mutual respect' which is to give room to substantive debate and conflict. They argue that our consensus on higher order principles should not serve to eliminate moral conflict from politics, but on the contrary, should regulate moral disagreement. In stead of 'principles of preclusion', we need 'principles of accommodation', in which we agree how to disagree on controversial issues. Unfortunately, liberal philosophers have given less attention to this way of dealing with principles (Gutman and Thompson, 1990, p. 124). In sum, the fundamental problem seems to be that liberal theory conceives of this consensus on higher order principles as antecedently defined and hypothetically agreed upon by the participating parties – thereby limiting the agenda – whereas Gutman and Thompson want this consensus to be strived for in the public debate, by the parties themselves.

Rawls' way of dealing with the democratic majority rule clearly illustrates how the just savings principle serves as a principle of preclusion as well:

> *I now wish to note the use of the procedure of majority rule as a way of achieving political settlement. As we have seen, majority rule is adopted as the most feasible way to realize certain ends antecedently defined by the principles of justice. Sometimes however these principles are not clear or define as to what they require. This is not always because the evidence is complicated and ambiguous, or difficult to survey and assess. The nature of the principles themselves may leave open a range of options rather than singling out any particular alternative. The rate of savings, for example, is specified only within certain limits; the main idea of the just savings principle is to exclude certain extremes* (Rawls, 1999, p. 318).

Here, we see that the just savings principle itself is regarded non-negotiable. Rawls explicitly narrows the scope of public debate on our responsibility for future generations to an almost trivial discussion on the exact interpretation of the just savings principle: the estimation of a just rate of savings. Thus, the antecedently defined principle itself and the definition of the problem are given a sovereign position, and kept beyond discussion. My criticism does not imply that liberal theory precludes or bans public debate on these principles, but rather, that it renders them redundant and meaningless. Liberal theory takes away the necessity to enter into debate about issues of environmental responsibility.

By privatising environmental concerns, liberal theory tends to neglect the structural nature and causes of the environmental crisis, for obviously, those collective structures are beyond the horizon of interest of the environmentally spirited consumer. The mainstream liberal answer implies that environmental problems can be solved within the current economic, cultural and technological structures of liberal capitalism by cultivating environmental awareness and stimulating citizens to behave according to principles of sustainability. Thus, the requirements of sustainability are generally assumed to be compatible with those of the economic growth paradigm and technological progress, underlying the structures within which people currently live and make their choices. Environmental policy is primarily understood as an economic strategy of resource distribution and management that aims to regulate our means of production and consumption in such a way as to overcome particular versions of the prisoners' dilemma that arise within these paradigms[10]. Likewise, environmental policy and education are regarded as means to politically predetermined ends. Moreover, the basic structure of society – our common economic, social and cultural institutions – is left untouched by this type of analysis. In my opinion, this focus on individual change in behaviour and mentality falls short of an adequate analysis of the environmental crisis as a social problem. Consequently the political dimensions of the environmental crisis are pushed into the background. Environmentalists and environmental theorists have argued convincingly that ecological awareness is not a private concern. Considering environmental interests in our daily behaviour- and consumption patterns is not only a matter of individual choice. On the contrary, our daily choices and behaviour are continuously driven and regulated by the prevailing economic and social structures in society. The Dutch philosopher Hans Achterhuis gave the label 'machine morality' (*moraal van de machine*) to these powerful mechanisms. As long as the environmentally harmful structures and institutions remain intact, Achterhuis argues, every moral imperative, every appeal to individual morality is useless, or even hypocritical (Achterhuis, 1993).

For example, our 'freedom of choice' as consumers is severely limited by the structure of the global food market (its modes of production, use of energy and 'raw material', its social policy, modes of transportation and trade). If consumers have serious objections against genetic engineering, for instance, they will find it very hard to stick to a diet of genetically 'sound' soy bean. This controversial ingredient is blended into the most commonly used products, without consumers being informed. Individual consumers hardly have any influence on the range of products that is offered to them, and their ecological quality. Therefore, the global food market leaves us little choice and gives rise to particular consumption patterns. And apart from this, central governments are no longer in a position to guarantee the basic quality of the products offered to us by the food market. Eventually, the environmental risks and risks of safety and health are beyond the grip of centralised institutions because every general effort to reduce risks creates risks on another level that are sometimes even harder to control. For example, in order to minimise the risk that our harvest will be eaten by parasites, farmers started using pesticides. Unfortunately, as Rachel Carson warned in her famous book *Silent Spring* (1962), those pesticides poisoned our groundwater and entered into our food chain, thus becoming a threat to ourselves as well as to other animals. In response to this risk, the agricultural industry started breeding parasite-resistant crops. However, the introduction of foreign and, eventually, genetically engineered crops together with the rise of mono-cultures created risks of ecological disturbance. In order to reduce these harmful effects on the ecosystem, governments founded expert institutions, designed to measure and control the use of crops and pesticides. Right now, those institutions run the risk of being suffocated by the immense bureaucracy and jurisdiction they engendered (Koelega, 2001). His insight in the dynamic processes by which centralist measures of risk control generate and disseminate even greater (social) risks lead Beck to describe late-modern society as a risk society. He argues that the standard responses to these risks – stronger state interventions – are no longer effective because they take away responsibilities from the public parties on the local level where those problems have to be dealt with (Beck, 1998). What follows from this brief analysis, is that environmental concerns are predominantly public concerns, which should be discussed and acted upon on a public–political level. Environmental responsibility should therefore not be restricted merely to the private sphere of consumers, nor to the public sphere of state control.

The sharp distinction between private and public spheres precludes an adequate approach to the environmental crisis. As I argued earlier, within liberal theory the private sphere is valued as the primary space where people are truly free to live according to their own conception of the good or life-plan. The public domain is only of derivative value. Public morality and politics are necessary to guarantee maximum freedom within this sphere. Consequently, the liberal concept of citizenship is ultimately minimal and negative. Citizenship is understood as a legal status, a complex of passive entitlements and some heavily circumscribed duties. Furthermore, there is no obligation to participate in the public life. That means, there is no incentive to engage in the vital practices of the civil society; the free space occupied by media, political fora, churches, unions, environmental groups, organisations of farmers, neighbourhood associations and so on, independent of the state, which unite citizens around issues of common concern (cf. Oldfield, 1990; Kymlicka and Norman, 1994, p. 353; WRR, 1994; Taylor, 1995; Pettit, 1997; Nauta, 2000). Within the public sphere no common ideals are to be pursued except for a shared conception of justice, which, as we saw, can neither represent the needs of future generations nor the intrinsic value of non-human nature. The liberal democratic state is committed only to protect the fundamental rights and interests of its present citizens. The impotence of contemporary Rawlsian liberalism to cope with the challenges of the environmental crisis is, in my opinion, rooted in this paralysing distinction between private and public spheres, and consequently in the minimal and negative conception of citizenship. Obviously, a more substantive and comprehensive understanding of politics and citizenship is needed.

Conclusion

In this section I have tried to reveal the inadequacies of the liberal framework that subtends the narrow understanding of sustainable development for environmental education. First and foremost, I have argued that Rawlsian liberal ethics cannot include obligations towards future generations because such generations are beyond the reach of our reciprocal relationships. However, in his recent paper *Creating Green Citizens? Political Liberalism and Environmental Education* (2004) Derek R. Bell argues that my understanding of both the nature and the place of Rawls's theory are 'flawed'. For Rawls, reciprocity is not the same as mutual dependency, so he argues. Instead, reciprocity is itself a political ideal:

> *A society regulated by the principle of reciprocity is one in which all citizens who do their part benefit fairly from their mutual cooperation. Citizens from non-overlapping generations cannot be mutually dependent but they can be engaged in a single cooperative venture, which for Rawls is 'realizing an preserving a just society' (...). For Rawls, there is no 'problem of reciprocity' because his notion of 'reciprocity' – as the fair allocation of benefits and burdens between those engaged in a*

> *cooperative scheme – is perfectly compatible with the idea of justice between generations* (Bell, 2004, p. 46).

In response, I would like to emphasise that Bell is right in stressing that Rawls – in some places – opts for a more broad interpretation of reciprocity between citizens in terms of 'equal fairness'. However, as I have previously argued, he cannot maintain this broad interpretation, because at another point in his magnificent theoretical construction Rawls makes the application of the principles of justice dependent on certain empirical conditions – the 'circumstances of justice'. As we have seen, the strong and wide appeal of Rawls' theory is mainly due to the fact that he derives strong obligations from weak assumptions concerning human nature. He does not assume a spark of altruism from the contracting parties. The 'circumstances of justice' are adopted in order to secure these weak assumptions. They outline the 'normal conditions' under which human cooperation is both possible and necessary, and thus, a conception of justice is needed. If these circumstances do not hold, or one of them does not hold, the principle of justice can find no application. Therefore, the 'circumstances' function as a constraint on the applicability of 'justice as fairness'. The 'circumstance of relative equality' requires that the parties 'are roughly similar in physical and mental powers; or at any rate, their capacities are comparable in that no one among them can dominate the rest. They are vulnerable to attack, and all are subject to having their plans blocked by the united force of others' (Rawls, 1999, p. 110). Obviously, our hypothetical relationship with future generations does not fit into this picture. Owing to our privileged position in time, we – contemporaries – can block the presumed plans or range of options of future generations. Furthermore, they cannot harm us in return. We are invulnerable towards future generations. Conversely, future generations are dependent on our capricious feelings of benevolence, or the coincidental convergence of their interests with our self-interest. Therefore, in my view, the principle of justice can find no application in an intergenerational context.

Of course, this negative conclusion concerns the *Rawlsian* liberal framework. As such, the scope of my criticism is limited to the Rawlsian discourse on intergenerational responsibility. Strictly speaking, I have not yet provided sufficient arguments that allow me to rule out the possibility of an *alternative liberal* framework that is fruitful with respect to the issue at hand, though I would not know any author in the field of liberal political theory who has provided a theory as well-defined and influential as Rawls'. While one should indeed be cautious when it comes to making generalisations on based on my conclusions concerning liberal theory in general, in the last part of this section I have put forward some arguments in support of my view that it will be very unlikely for any liberal political theory to deal with the issue of future responsibility in a satisfactory way. First, I have shown that a contractual understanding of justice is less neutral than its advocates want us to believe. In fact, the contractual notion of intergenerational responsibility – captured by the narrow understanding of sustainable development – posits an economic-distributive and anthropocentric perspective on the issue of environmental responsibility. This perspective leaves intact the problematic underlying structures of society within which citizens make their choices and their behaviour is being regulated.

Second, liberal theory privileges autonomous citizenship in a way that places the self at a distance from his sources of meaning and identification. As such, it is likely to assume that the incentives inherent to the contract will elicit an instrumental rather than an intrinsic involvement in public affairs. The sharp distinction between private and public concerns easily leads to an impoverished practice of environmental morality and politics. Furthermore, in contrast to liberal minimalism, the creation of a shared environmental morality requires intense political debate and struggle. It requires debate and struggle about such moral questions as: what does the future mean to us? What kind of world do we want for our children and grandchildren? Which sacrifices are we prepared to make now in order to realise this? I have argued that these questions can no longer be restricted to the private sphere but have to be confronted in the public arena of environmental politics and education. My criticism of the liberal framework of intergenerational responsibility will be elaborated in section 2.3.

The meaningfulness of these questions itself reveals that there must be some other kind of reciprocity between us and those unknown descendants, who will live centuries from now, albeit that this is not a reciprocity in any strict, procedural sense. We need a sense of the future in order to make our life here and now worthwhile. We need a sense of continuity – a sense that our worries, hopes and dreams have not been in vain because life on earth will continue – and our worldly activities and projects will be carried on. Maybe we – reflexive creatures that we are – need the idea that there will be future generations to remember us, just as we remember – and sometimes praise – past generations for who they were and what they did (for us). In short, the idea of a virtual bond with past and future generations seems to be an anthropological *sine qua non* for humans to make sense of their life. From this perspective, our moral obligations towards future generations might turn out to be less free-floating than the assumptions of procedural contractualism suggest. For one thing is clear, the liberal notion of intergenerational justice seems to present us with a false dichotomy. Either it leads to a wishy-washy domestication of environmentalism – treating ecological values and standards as mere lifestyle options – or it ends in practices of indoctrination – imposing sustainable development as a firm ideology upon (future) citizens.

In my view, a proper justification of intergenerational responsibility should not only be logically consistent, but should simultaneously articulate our personal commitment to a common future. Obviously, liberal theory not only fails to meet the first condition but the latter as well, because in my eyes it abstracts too much from the things we care for, here and now. The communitarian analysis of intergenerational responsibility – which will be central in the next section – starts with these communal sources of meaning and value.

2.2 FUTURE GENERATIONS AS HEIRS OF OUR COMMUNITY

That which binds us together as a society is not merely a common understanding of justice, but a common history, a common culture, morality and identity as well, thus communitarian critics stress in response to the liberal focus on rights and justice between individuals. Consequently, they argue, our obligations to future generations should not be regarded as exclusively determined by our individual allegiance to a contract of justice. Primarily, intergenerational duties arise from our commitment to the communal practices in which we are involved, and our commitment to the continuity of those practices. We take responsibility for future generations because they will be the heirs of our community, and as such, they represent our hope and trust in a prosperous future for our community. They will have to carry on our communal traditions, projects and ideals after our own death. To sum up, it is not our allegiance to the community of justice but the solidarity we experience within our moral community, which obliges us to take care for future generations.

Therefore, communitarians do not locate the source of our responsibilities beyond our 'selves' – within the claims of future generations – but within our present identity as community members. The objects of our present strivings do not cease to exist with our lives. On the contrary, our commitments reach beyond our own demise into the indeterminate future. John Passmore, for instance, notes on his commitment to philosophy 'To love philosophy – to philosophise with joy – is to care about its future as a form of activity: to maintain that what happens to it after our death is of no consequence would be a clear indication that our "love of philosophy" is nothing more than a form of self-love' (Passmore, 1974, p. 88). Inherent to our membership of a moral community is a sense of continuity. We participate, so that those activities we care about continue to exist. Without basic trust in the continuity of our projects and activities after our death, most if not all present efforts would lose their meaning. Communitarians therefore start with the things we care about here and now. Furthermore, they examine what kind of responsibilities human beings take up by caring for something near and dear.

The most detailed communitarian theory of intergenerational responsibility is presented by the Israeli philosopher Avner de-Shalit in his influential book *Why Posterity Matters* (1995). In line with the intuitions previously expressed, De-Shalit derives strong obligations to future generations from our sense of belonging to a community that stretches out over several generations, i.e. from the membership of a transgenerational community. The life span of a community covers a longer period than the life span of one of its members. Unlike Rawls' contractual understanding of the community of justice, membership of this community is not contingent upon personal consent or choice. We cannot swap communities – leave communities and enter others – like we change clothes. My ties to the community are not external to my self, but on the contrary, they define me. These ties and commitments are in a sense constitutive for personal identity. As such, de-Shalits understanding of the relationship between individual and community corresponds with the well-known communitarian thesis of the 'embedded self', as expressed by Sandel, Taylor, MacIntyre and many others. De-Shalit expresses this thesis by recalling the Aristotelian insight that 'he who lives outside the community is either too good or too bad, either subhuman or superhuman, or in other words, non-human' (de-Shalit, 1995, p. 15). Specific to his theory, however, is the thesis that we are embedded in a community that covers more than those generations with whom we live together; a sense of connectedness with those members who gave shape to our community as we inherited it, and those members who will carry on our heritage into the future. Against this background, de-Shalit criticises liberal theory for abstracting the individual from the community he is bound to, and thereby, cutting him off from the personal sources of meaning and inspiration. Second-order values, like impartiality and tolerance, will not be strong enough to motivate the kind of sacrifice and effort that an adequate approach to the environmental crisis will require of us.

The transgenerational community and moral similarity

Whereas de-Shalit appeals to strong and convincing examples of transgenerational communities – ranging from the Tibetan Buddhists to the British Labour Party – he experiences huge difficulties in his effort to define what precisely constitutes an intergenerational community. As his argument proceeds, de-Shalit comes up with three necessary characteristics of a transgenerational community. If these characteristics do not hold in the intergenerational relationship, then there is insufficient ground for obligations to past and future generations. First, there is the characteristic of cultural interaction, second, the characteristic of moral similarity and, third, the characteristic of reflection. These conditions are closely interrelated, and I will address them as such, since one follows from the other. For instance, de-Shalit argues that cultural interaction between people presupposes a certain degree of moral similarity; we need a common stock of values that functions as a moral background for conversation: 'In every genuine community some values and some attitudes towards moral and political questions are common to most people and serve as a background or as a framework when the members engage in discourse on their political and social life' (de-Shalit, 1995, pp. 27–28). A genuine community, de-Shalit argues, is one whose members are in search of moral similarity. However, the conditions of interaction and moral similarity appear to be extremely problematic in our relationship to future generations, as Martin Golding argued in one of the first philosophical papers on this issue, titled *Justice between Generations* (1972).

In Goldings' view – which we would now label communitarian – the existence of a social community is of decisive importance for the establishment of obligations. We measure our obligations by reference to our relationship to those who make a demand on us. Whether the claim of the other will be recognised by me, and whether I ought to feel obliged to respond to his or her claim, depends on the existence of a moral relationship. Knowing the values and ideals of the other is necessary, though not enough, Golding argues, for I could still simply ignore or reject his conception of the good. For the establishment of a moral relationship it is necessary that I recognise that 'his good is good to me'. That is, I should count his 'good' to be mine as well. Both parties have to share a common good. Only if I acknowledge his good as mine, can I feel obliged to contribute to the conditions of his good life. For this reason, our obligations to strangers are never immediately clear from the start. Suppose, for instance, that we were visited by creatures from a far-away planet, say Mars. In order to lay claim to our achievements and share in our wealth, Golding expects those strangers to emphasise their moral similarity to us: 'your good is our good'. Initially, we would probably reject their claims because 'they are not like us'. However, as the interactions between them and us increase, we will gradually be more inclined to recognise them as 'one of us'. And consequently, we will be inclined to recognise their claims and will take our obligations to them more seriously. In a similar way, our history reveals a gradual extension of civil rights to labourers, women and children (Golding, 1972).

What are the consequences of Goldings' analysis for the establishment of obligations to future generations? Whether we should feel obliged to contribute to the well-being of future generations depends on the possibility of sharing a common good. Obviously, we do not share a common life with future people. Therefore, we lack the opportunity of interaction with future people, and *ipso facto*, we lack the opportunity of developing a common ideal or conception of the good. Their good is not necessarily good to us, and our good is not necessarily good to them. For this reason, Golding draws a rather pessimistic conclusion: 'It appears to me that the more remote the members of this community are, the more problematic our obligations to them become. That they are members of our moral community is highly doubtful, for we probably do not know what to desire for them' (Golding, 1972, p. 97). Therefore Golding argues that since there is no common moral framework we can rely on, we should keep our hands off their good life.

In general, *face-to-face* interaction is assumed to be a necessary characteristic of a moral community. Though de-Shalit agrees with Golding that intergenerational obligations should be grounded in a common conception of the good, and though he agrees that a certain kind of interaction between the moral parties is a preresquite for the establishment of such a common good, de-Shalit does not believe that this interaction should necessarily be a *face-to-face* interaction. The obvious impossibility of direct communication between people, who do not exist simultaneously, does not preclude other kinds of cultural interaction between and across generations. In our human coexistence as cultural and historical beings there are many indications to be found for intergenerational interaction. De-Shalit, for instance,points to the interaction embodied in a common canon of artworks, evolving traditions, stories and rituals. By engaging in a common culture, by revaluating our great works of art, by retelling our stories and carrying on our traditions and rituals in our own way, *we are in conversation* with our ancestors, and we are inviting future generations to respond and carry on. Every cultural act inserts itself into a historically evolved tradition and alludes to some future apprehension. In my view, this profound insight is most eloquently expressed in the hermeneutics of Hans-Georg Gadamer. Gadamer reveals that if we are to make sense of ancient works of art, if we want those works to 'speak' to us, a fusion of horizons ought to be established between the ancient world of the artwork and ours. Such a fusion of horizons into a common horizon of significance cannot be established directly, because we are connected to those ancient works of art by means of an 'effective history' (*eine Wirkungsgeschichte*), that is, a trace of interpretations that its reception left behind in the course of history. Only by following this conversation of the intergenerational community of reception throughout the years that divide us, we will be able to bridge the time-gap in between us and the ancient work of art (Gadamer, 1960).

So, apart from the face-to-face interaction that we enjoy as contemporaries, there is a kind of genuine interaction between generations and continuity across generations that allows us to speak of a transgenerational community. This is most obvious, when looking at the history of moral communities from the outside: 'From the perspective of the outsider, the community remains constant over generations although its members change; people die and others are born, but we continue to speak of the same community: the British Labour party, the Tibetan Buddhists, etc' (de-Shalit, 1995, p. 21). More problematic and controversial however is de-Shalits thesis that underlying this interaction and continuity is a search for moral similarity: 'Cultural interaction, then, exists when, in addition to the common language, codes, and tradition of symbols, there is a cultural, moral and political debate. That is when people accept that they are governed by common values and principles' (de-Shalit, 1995, p. 25). What is unclear is what de-Shalit precisely means by postulating 'a search for moral similarity'. Obviously, de-Shalit believes communities organise themselves around shared moral and political values, and these shared values form an interpretative framework. He uses the metaphor of 'moral spectacles' to indicate the function of these shared values as a backdrop for political debate.

However, it still remains unclear as to what the status and 'thickness' of this moral similarity are. In his careful search for a profound position, de-Shalit finds himself in between two positions to which he is not ready to subscribe. First, the moral similarity he has in mind is not fixed or given beforehand by some religious authority or metaphysical doctrine, as some conservative communitarians seem to suggest. De-Shalit is reluctant towards thick notions of community which are thought to determine our moral values. Moreover, de-Shalit argues that the notion of a moral community should be compatible with the notion of the 'free and rational agent'. That means that, in his view, membership should be self-chosen, members should reflect on their membership and the values embodied in their community practices. Furthermore, the community should remain open to debate and change. If these conditions do not hold, de-Shalit argues, there can be no transgenerational community in a genuine communitarian sense (de-Shalit, 1995, p. 16–17). So, the moral similarity he proposes is less thick, and its status is less solid than the moral similarity of many fellow communitarians. On the other hand the moral similarity de-Shalit has in mind is thicker than that which the liberal contract allows since it includes a shared self-understanding. And the common good encompasses comprehensive cultural and moral values that would obviously conflict with the liberal demand of impartiality and neutrality. However, the moral similarity de-Shalit postulates, is of an empirical rather than a Rawlsian transcendental nature. The change of values is more like a cultural evolution or the expression of an ongoing conversation, than a result of procedural deliberation.

Obligations to immediate and remote posterity

The concept of a transgenerational community enables de-Shalit to establish strong obligations for future generations that go beyond the minimal obligations of distributive justice. After all, we should regard future descendants as the heirs of our community with whom we share a common conception of the good life. And because of this commonality we are able to judge what is good for them and enjoy a moral relationship. In de-Shalits own words: 'In this theory, one's self-awareness is related to one's community, both in the present and in the future, i.e. in relation to the aims, desires, ideas, dreams and values of the transgenerational community. By extending the community to include future generations, I have argued that obligations are owed directly to them, and that since these obligations are owed directly to them, and since these obligations derive from the community that constitutes our "selves", contemporaries should take these obligations very seriously, as they indeed have good reasons to do' (de-Shalit, 1995, p. 124). Thus, communitarian theory appears to liberate us from the problems of neutrality, paternalism and ignorance. Unlike Rawls' contract community, the transgenerational community provides us with a base for strong, positive and negative obligations to future generations: 'That is, we should consider them when deciding on environmental policies; we should not overburden them; furthermore we should supply them with goods, especially those goods that we believe are and will be necessary to cope with the challenges of life, as well as other, more non-essential goods' (de-Shalit, 1995, p. 13).

However, since the transgenerational community is not immune to change, its life-span is not unlimited. De-Shalit is ready to admit that the transgenerational community does not reach into the indefinite future because our moral similarity with future generations will diminish with the passage of time. Inevitably, there will be a particular moment in the future, when we no longer recognise their good as ours. In the evolution of practices and traditions, the embedded goods will gradually transform into goods that are too far removed from the goods that once made our lives worthwhile. Consequently, we will no longer feel obliged and motivated to contribute to the conditions of their good life. At that particular moment in the future, our membership of the transgenerational community comes to an end, and our obligations fade away. De-Shalit stresses that we should not want to have it otherwise, if we take the normative requirement of a genuine transgenerational community seriously: 'We can reasonably predict that changes in technology, together with other factors, will lead to changes in the values held by future people. What is more, we would like this to happen, or at least to know that reflection will take place and that if future generations agree with our norms, values and policies, it is because they find them good and not because they have not reflected upon them. Ultimately, we would rather be sure that our transgenerational community is a genuine constitutive community than see our values forever accepted without a reasoning process taking place' (de-Shalit, 1995, p. 50).

When future generations are no longer part of our community, our strong obligations towards them vanish because there is no moral similarity between them and us anymore. For this reason, de-Shalit introduces a sharp distinction between our obligations to immediate posterity and those to remote posterity. We have strong, positive *and* negative obligations towards those future generations closer to us. That means that we are in a sense, co-responsible for their realisation of the good life. We should consider their good as ours and contribute to the conditions of their good life: '(…) we should supply them with goods, especially those goods that we believe are and will be necessary to cope with the challenges of life, as well as other, more non-essential goods' (de-Shalit, 1995, p. 13). As time goes by, these strong obligations fade away until they have completely vanished. But this does not imply that we do not owe them anything, that we do not have to consider their needs in any way. According to de-Shalit we still have a negative duty to prevent 'severe predictable harm' to the life conditions of remote posterity: 'To people of the very remote future we have a strong 'negative' obligation – namely, to avoid causing them enormous harm or bringing them death, and try to relieve any potential and foreseeable distress' (de-Shalit, 1995, p. 13). Our obligations to those future generations are similar to our obligations to strangers. These negative obligations cannot be derived from a common good but are grounded on universal principles of humanity. With a reference to Rawls' theory, de-Shalit notes that: 'This is a matter of humanity rather than of justice. The difference between the two is that justice is concerned with principles of ownership or the control of resources, while humanity is concerned with people's well-being. It requires us to avoid the infliction of suffering and to relieve it where it occurs' (de-Shalit, 1995, p. 63). The split between strong communitarian and weak humanitarian obligations corresponds closely to our common sense intuitions, de-Shalit argues. We are more prepared to contribute to the good life of those future people near and dear to us than for those far-away people we can hardly picture, although we would not want to harm either one (de-Shalit, 1995, p. 13).

Evaluation

Where does this communitarian approach lead us to? By taking the powerful concept of a transgenerational community as a ground for justification of the establishment of duties to future generations, de-Shalit ingeniously dissolves the four problems of justification that liberal contractualists see themselves confronted with (see section 2.1). First, the problem of reciprocity does not arise, because our obligation to consider the good of future generations is not dependent on a situation of mutual dependency in which the parties are assumed to be motivated to cooperate out of mutual self-interest. Moreover, it is because of the good life we share that we are intrinsically motivated to respect one another. Second, de-Shalit has overcome the problem of ignorance by understanding future generations as fellow-members of our community who are very much like us. We know what to wish for them, because their good is ours as well. Choices about what to sustain for them are made on grounds of a shared conception of the good. Third, the problem of paternalism does not hold, for, according to communitarian theory, intergenerational relationships are in a sense paternalistic by nature. In a sense, paternalism can be seen as a function of the intergenerational transmission of community practices and traditions. That means, we act out of their interest by acting out of the interest of our community. Fourth, communitarian theory liberates us from the difficulties of the non-identity problem, since its thesis does not depend on the claims of those who will be personally affected by our present action, but on our striving for continuation of our communal practices.

For this reason, de-Shalit does not have to employ a gloomy distinction between potential and actual persons (Feinberg, 1974b; cf. de-Shalit, 1995, p. 125).

It is indeed fascinating to see how these four problems of justification lose their significance in a communitarian framework, but more important is de-Shalit's answer to the problem of motivation, as I stipulated earlier. How are we assumed to be motivated to do justice to posterity? According to communitarians, our desire to secure a liveable world for future generations follows naturally from the fulfilment we find in our lives, here and now. In fact, de-Shalit locates the source of intergenerational obligations in our present commitments to those values and ideals embedded in our community practices that constitute our identity. That is, a genuine self-awareness obliges us to consider the good life of future members of our community.

Another advantage of the communitarian view of intergenerational responsibility is the acknowledgement that the environmental crisis comprises a complex of collective problems that requires collective responsibilities rather than individual choices. We have to act, not because our individual rights, needs or properties are in danger. Moreover, it is because our 'common goods' are being affected or even threatened with extinction as a result of collective behaviour. This collective threat requires us to determine which goods constitute our community life and should be preserved for the heirs of our community. Understood in this way, the common good is more than a mere sum of individual preferences or a convergence of private interests, as some liberal theorists suggest. The whole is more than the sum of its parts, since the common good is constituted in public discourse or to be more precise, it is mediated by common practices in which we come to recognise our good in the eyes of others. More than a passive consent of individual parties, the common good requires an active identification and commitment of the community that sustains it (cf. Taylor, 1995, pp. 31–34). Therefore, communitarian theory allows for a more comprehensive practice of citizenship. The sustaining of our common good calls for an active participation in communal practices and a personal responsibility for community life.

Unfortunately, there are fundamental problems, apart from these strengths, that are also connected to this communitarian model of intergenerational justice. The most obvious weakness is its limited scope of justification, in temporal as well as 'social distance'. To start with the latter: since de-Shalit derives obligations from the goods that bind us together as community members, neither do our obligations reach beyond the limits of our moral community. Therefore it is unclear how communitarian obligations are to contribute to an effective approach of environmental problems of a global scale. De-Shalit might respond that everyone should care for its own community-members, and as such, every single person on this globe is being cared for. However, on this particular point, the idea of a strong transgenerational community rests on a more or less outdated view of society. De-Shalit appeals to the picture of a stable society neatly compartmentalised into coherent communities, along religious and ideological lines. Its members are assumed to live and die within the same community and are assumed to devote their lives to the ideals and projects of this particular community. However, this picture is in sharp contrast with the reality of late- or post-modern society, profoundly marked by plurality and a general shift from a standard biography to a choice biography. Sociologists argue that individual lifestyles take precedence, that internally coherent communities are being threatened with extinction and our common values as a society are becoming subject to erosion. With this loss of unity, an appeal to community spirit loses its grip.

The problem of limited inclusiveness repeats itself in the time dimension: it remains unclear as well how far communitarian obligations reach into the future. Does our collective responsibility cover a time span of three generations, or perhaps ten or twenty? De-Shalit is ready to admit that the transgenerational community is limited in time, but he veils himself in rather vague terms when it comes down to concrete questions with an empirical bearing, such as: how are we to determine – with only present knowledge available to us – at which particular moment in the future our community stops being *our* community? Here the problem of ignorance rears its ugly head again, albeit in a different way.

Finally, and in connection with the latter, there are fundamental objections that should be made against the expression of community commitments and solidarity in terms of abstract values, in particular, in situations where communities claim exclusive ownership of those values, as if it were possible to patent something like 'genuine respect for posterity'. This objection is most clearly articulated by Blake, Smeyers, Smith and Standish: '(...) it is arguably a failing of communitarianism to characterise this interconnectivity in abstract terms of values and principles, rather than in terms of interlocking and sometimes mutually interfering narratives. In emphasising supposedly shared abstractions rather than the concrete projects of making sense, in and amongst each others' lives, communitarianism risks becoming yet one more variety of negative nihilism, positing some kind of transcendent sense of the good, over against meaning and value immanently constructed by ourselves' (Blake et al., 2000, p. 45). Thus, communitarian theory cannot come to grips with the reality of late- or post-modern citizens. They no longer enjoy an all-embracing membership of one community, but participate in varying associations and practices. And from their disparate commitments they extract disparate roles, identities and loyalties, which sometimes conflict with one another (Van Gunsteren, 1992, pp. V–VI; cf. Mouffe, 2000, p. 59). As such, de-Shalit seems to adopt a rather static concept of personal identity. In other words, de-Shalit glorifies the idea of a pre-political community (in contrast to the liberal glorification of a pre-political individual) but fails to recognise the reality and power of political community or the civil society. In the next section I will argue that an intergenerational ethic should take its bearings from this powerful public sphere, independent of the private world of the family and moral community on the one hand, but independent of direct state intervention and public institutions on the other.

2.3 FUTURE GENERATIONS AS IMAGINED STRANGERS

Until now both liberals and communitarians have failed to give an adequate account of our responsibility for future generations. Whereas Rawls' attempt to include future citizens in the contract community fails because of the absence of reciprocity – understood as mutual dependence – between people of different generations, de-Shalit assumes a substantive communal reciprocity that is too limited of scope and, even in its limited form, no longer present in late-modern society. The shortcomings of both perspectives are in my view due to the fact that they abstract too much from the actual political community that binds us together as citizens, namely the civil society. The civil society defines the intermediate, public sphere of voluntary associations between citizens, beyond the private sphere of the family and moral community, but also independent of immediate state intervention. This is the sphere where public opinion on environmental issues is raised, where debate about the institutionalisation of ecological responsibility takes place, where parties hold one another responsible for their involvement with the natural environment, and where the public agenda is settled. Moreover, this is the sphere where people identify themselves as political actors and where they engage in collective action and take responsibility for the public good. In this section I will draw on the ideas of traditional civil republicans, such as Hannah Arendt and Adrian Oldfield, as well as on the ideas of new representatives within this progressive school of thought, some of whom call themselves neo-republicans, others as *civil society theorists*, and still others as advocates of a radical democracy (as distinguished from the ideal of a deliberative democracy in the Habermasian or Rawlsian tradition), such as Chantal Mouffe. Altogether, those authors have inspired me to develop an alternative account of citizenship and future responsibility; an account that I will refer to as a neo-republican account. I will present my alternative account of citizenship, by contrasting my ideas and arguments with those of liberal and communitarian theory.

The civil society and its citizens

Within liberal theory the private sphere is valued as the primary space where people are truly free to live according to their own conception of the good life. The public domain is only of a derivative value: public morality and politics are necessary to guarantee maximum freedom within this sphere. Consequently, the liberal concept of citizenship is ultimately minimal and negative. Citizenship is understood as a legal status, a complex of passive entitlements and some heavily circumscribed duties. Furthermore, there is no active responsibility or obligation to participate in public life. Within the public sphere no common ideals are to be pursued except for a shared conception of justice, which, as we saw, can neither include future generations nor non-human nature. The impotence of contemporary liberalism to cope with the challenges of the environmental crisis is, in my opinion, rooted in this paralysing distinction between private and public spheres, and consequently in the minimal and negative conception of citizenship. Obviously, a more substantive and comprehensive understanding of environmental politics is needed. In their response to liberal minimalism, communitarians have succeeded in locating the communal sources that give meaning to our life and future life. Moreover, they promote an active participation in community practices. However, in my view, communitarians altogether fail to formulate an adequate alternative for the liberal contract doctrine, because the participation and common commitment they require is limited to the private sphere as well, albeit not the private sphere of the individual, but that of the moral community. Therefore, in my opinion, communitarian theory is not so much about citizenship as it is about about community membership (and as such perfectly compatible with the liberal ideal of citizenship). Insofar as communitarians do address issues of citizenship, they tend to speak to citizens as spokespersons of their community. In this sense, communitarians reconfirm the liberal hierarchy between private and public spheres. Whereas liberalism resorts to a pre-political private subject, communitarianism embraces the idea of a pre-political community.

Following the neo-republican analysis along general lines, I would like to argue that a viable democracy, one which is strong enough to deal with the societal problems arising from the environmental crisis, is in need of a more comprehensive and substantive theory and practice of politics. Obviously, a revaluation is needed of the traditional liberal hierarchy between private and public spheres. A reversal of this hierarchy is commonly proposed within the neo-republican tradition of political thought, expressed by manifold philosophers throughout western history, moving from ancient Greece (Aristotle) and Rome (Cicero) to Machiavelli and modern philosophers like Hegel, Rousseau and Arendt. In contrast to the liberal glorification of private life, the main emphasis of neo-republicanism is on the intrinsic value of political life. According to some, political life is even superior to 'the merely private pleasures' of family, community and neighbourhood, dictated as they are by the biological necessities of human consumption and reproduction. In this view, political action comprises 'the highest form of human living together' (Arendt, 1958a; Oldfield, 1990, p. 6; Kymlicka and Norman, 1994, p. 362; Honohan, 2002).

According to Hannah Arendt, for instance, human destiny lies on the public stage; the highest human capacities, those of public speech and action, are realised in political debate and struggle and find their home in the public arena. Only here, are human beings able to live together as free and equal, though radically different beings. It is within our 'intersubjective web of human relationships' that individual words and acts acquire a public meaning. Simultaneously, in public speech and action a glimpse of our personal 'self' is being expressed. In public action we expose ourselves, we display who we are and what we stand for. However, our words and actions do not express a pre-given essence or identity. On the contrary, our self is constituted by the public expression itself; we create ourselves anew in words and actions by enacting our life stories in a public web of narratives, meanings and relationships. In this view, the activities of the citizen are premised on the demands of freedom, whereas the labour of the *animal laborans* – the labourer – is dictated by the biological necessities of life, and the work of the *homo faber* – the working man – is controlled by a compulsive instrumentality (Arendt, 1958b).

There is, however, a difference between this neo-Athenian tradition of thinking about citizenship, that elevates political activity to the rank of a supreme end in human life, and the neo-Roman tradition in which our civil activities are regarded intrinsically valuable among other activities we engage in as family members, professionals, community members, lovers and so on. In this latter view, our civil responsibilities do not necessarily transcend other responsibilities and loyalties.

However, in contrast to the liberal conception of *citizenship-as-legal-status*, both neo-republican sources value citizenship as a desirable activity, an active responsibility shared with fellow-citizens. In this view, citizens should engage in common practices and debates on the problems they find themselves confronted with. Insofar as one can speak of a common morality, purpose, good or ideal, these are not given beforehand. Rather, these goods and ideals are continuously defined, redefined and evoked in the political debate and struggle itself (cf. Oldfield, 1990; Kymlicka and Norman, 1994, p. 353; WRR, 1994; Taylor, 1995; Pettit, 1997; Nauta, 2000).

Iseult Honohan offers a definition of neo-republicanism that clearly distinguishes this school of political thought from that of liberalism and communitarianism (while acknowledging that there are also similarities among these schools of thought):

> *Civic republicanism addresses the problem of freedom among human beings who are necessarily interdependent. As a response it proposes that freedom, political and personal, may be realized through membership of a political community in which those who are mutually vulnerable and share a common fate may jointly be able to exercise some collective direction over their life (...) In this approach, freedom is related to participation in self-government and concern for the common good (...). Emphasising responsibility for common goods sets republicanism apart from libertarian theories centred on individual rights. Emphasising that these common goods are politically realised sets republicanism apart from neutralist liberal theories which exclude substantive questions of values and the good life from politics. Finally, emphasising the political construction of the political community distinguishes republicans from those communitarians who see politics as expressing the pre-political shared values of a community* (Honohan, 2002, p. 1).

As opposed to the mainstream liberal conception of a common good as an addition of convergent goods that can be defined prior to public discourse, Honohan states that the common good is constituted by public discourse itself: 'The model of common good central to republican politics is that of intersubjective recognition in the joint practices of self-government by citizens who share certain concerns deriving from their common vulnerability' (Honohan, 2002, p. 156). And elsewhere:

> *The common good is realised in the activities of participants for whom membership in the community of the practice is part of living a worthwhile life (...). The common good towards which members are oriented is the flourishing of those practices, and this depends on the quality of participation of members. Thus there are common goods which are not decomposable into individually distributed goods, and cannot be understood wholly instrumentally. These goods are neither a property of the whole, nor determined by the goal of an organic entity. Thus, in this context, it is more appropriate to speak of common goods than "the" unitary common good* (Honohan, 2002, pp. 153–154).

Common goods, as distinct from convergent ones, depend on common meanings and practices that are socially defined and closely interconnected. You cannot, for example, be a vegetarian in a society of meat-eaters that is unfamiliar with vegetarianism. Others will simply fail to recognise the reasons the vegetarian expresses for not eating meat: he will either be declared insane or he will be looked at as someone who does not enjoy meat or who is apprehensive about eating dead animals. Moreover, another constituent part of republican thought is the idea that sometimes common goods have a certain priority over individual goods, precisely 'because they consist of practices and possibilities through which individuals realise themselves in many dimensions essential to human fulfilment. But common goods should not be thought of as inherently in conflict with the good of individuals, but as part of the good of individuals. Nor are they essentially in tension with freedom, if it is understood in terms of autonomy. Autonomy is a matter of acting according to one's most significant purposes, and needs some social framework to support' (Honohan, 2002, pp. 153–154).

Neo-republicans argue that the requirements of autonomy are not mainly of an individual kind, as liberal theory suggests. Moreover, the possibility of political autonomy implies collective action. Since we read off our sense of who we are from the social world we live in, much of our autonomy can only be won by changing the social world collectively (Smith, 1998b). Couched in terms of Isiah Berlin, the neo-republican notion of autonomy rests on a more positive concept of freedom in terms of self-government, self-expression and participation, rather than on a negative concept of freedom in terms of non-interference and the absence of obstacles to the fulfilment of the individuals' desires. Furthermore, the requirements of political autonomy are by no means only of a rational kind, and they are only loosely related to neo-liberal notions of choice, needs and wants. It is not so much by means of rational choice but by means of collective action and meaningful practices that we deal with the problems arising from our distorted interaction with the natural environment. Thus, we strive to free ourselves from the structural constraints that restrict our freedom (Dewey, 1916; Berlin, 1958; Oldfield, 1990; Pettit, 1997; Smith, 1998b).

For neo-republicans autonomy is the prerogative of those who act in a public space. As such, the status of autonomy and citizenship is contingent upon this activity. Only of those who engage in political action and make judgements about common concerns in dialogue with fellow citizens can be said that they enjoy political autonomy. Consequently, individuals can lose political autonomy, if they fail to take up their responsibility as a citizen. This is one of the reasons why Oldfield describes neo-republicanism as a 'hard school of thought'.

> *One can cease to be a citizen if one no longer performs the duties, but this does not mean that one loses the 'rights' – for the rights, contra the liberal-individualist position, belong not to citizens but to individuals (...). Not to perform the duties of citizenship, therefore, is not to lose one's rights, but to lose the esteem of one's fellows. It is to declare that*

> *you prefer them to take on the responsibility of politics; it is to abdicate from self-government; it is to cease to be a citizen. But it is not to lose the possibility of becoming a citizen again* (Oldfield, 1990, pp. 160–161).

Civil republicans, neo-republicans, civil society theorists and representatives of a radical democracy have in common that they strive towards a re-politicisation of society. They reclaim a free, public space for active citizens against the expanding powers of the market economy, and against the pervasive interference of the state. Such an intermediate space, in between the private and public sphere, is marked by the civil society. According to Charles Taylor, a civil society should be conceived of as 'a network of institutions that are independent from the state, which unite citizens around issues of common concern and which by their very existence, or by their activities, can bring influence to bear on policy' (Taylor, 1995, p. x; cf. Kymlicka and Wayne, 1994, p. 363)[11]. Churches, unions, ethnic associations, co-operatives, environmental groups, neighbourhood associations, women's support groups and charities are among those institutions or voluntary associations at the core of the civil society. Within the neo-republican school of thought – civil society theory in particular – heavy weight is placed on everyone's active participation in these voluntary associations. Through their voluntary participation and association citizens are expected to develop a public spirit and a common commitment to issues of a public interest. Perhaps one of the strongest expressions of a civil society is the proliferation of public opinions through the widespread debates, disseminated by all kind of media, forums and public institutions. What was particularly unique about the emergence of a public opinion in the nineteenth century was that it arose from sources that, for the most part, were wholly independent of church and state authority. Historically, this indicates the rise of a critical public or critical mass (Taylor, 1995).

Rather than being some kind of abstract ideal, a global civil society is actually emerging within current environmental politics. The state is no longer in the supreme position of a top–down director or arbiter as in the heydays of the welfare state. There are indeed more players in the field, as the rise of a global movement of grassroot resistance – initially labelled the anti-globalisation movement – indicates. Thus, environmental policies are continuously being shaped and reshaped in the political struggle between environmental activists, NIMBY groups[12], governments, trade unions, networks of biological farmers, consumers' organisations, international institutions like the World Trade Organisation, industry and car lobbies. Considered from a historical perspective, environmental politics is initiated neither by politicians nor by policy-makers. Environmental issues are placed on the political agenda by environmental activists and lobby groups of alarmed citizens (Beck, 1992; Waks, 1996; Klein, 2000; Hertz, 2001).

The emergence of a powerful global protest movement reveals a development that Beck describes as the 'sub-politicisation' of society: a shift of political power from the traditional political centre, marked by official state institutions of representative democracy towards the civil society, or what Beck labels the domain of 'sub-politics' (Beck, 1992). This shifting focus of the critical mass is not without reason. After all, national representative democracies lose power in favour of international institutions like the European Union, the United Nations, the World Trade Organisation, the World Bank and the International Monetary Fund. The most important global issues, ranging from regulations on biotechnology and genetic engineering to agreements on energy reduction, the conditions of livestock transport, water management and corporate investments, are decided upon by these global institutions. Simultaneously, democratic institutions lent their ears to multinational corporations that enforce free trade zones – free from social and environmental regulations. Moreover, former government responsibilities and public services concerning public transport and water supply, for instance, are taken over by private corporations. The international protest movement emerged as a means of resistance to these new centres of political power, that are either democratically weak or have no democratic legitimacy at all. Paradoxically, at the world conferences where these parties gather, the main focus is not merely on what happens inside the conference buildings but on the protests on the streets outside. This shift of focus indicates the dialectical power of the global civil society movements. In her popular book *The Silent Takeover* (2001) Noreena Hertz expresses the general cynicism and indignation that motivates people to join the protest:

> *As we have seen, while the power and independence of governments withers and corporations take over ever more control, a new political movement is beginning to emerge. Rooted in protest, its advocates are not bounded by national geography, a shared culture or history, and its members comprise a veritable ragtag of by now millions, made up by NGO's, grassroot movements, campaigning corporations, and individuals. Their concerns, while disparate, share a common assumption: that the people's interests have been taken over by other interests viewed as more fundamental than their own – that the public interest has lost out to a corporate one (...). The apparent inability or unwillingness of our elected representatives to defend our interests against those of business has created a cycle of cynicism (...). While some may welcome the recent attempts of various corporations to address some of the failings of the system and contribute to the social sphere, they tend to see these attempts as window-dressing or corporate PR, and remain sceptical about companies' motives. At the same time, they reject representative government as an ineffective, coopted and flawed mechanism for dealing with the failings of the market or representing their interests on the global stage, and reject the politics of today as the 'politics of Narcissus', concerned only with presentation and 'spin'. They choose to voice their concerns on the street, on the Internet, and in the shopping malls, because they feel that these are the only places that they can be heard. They will not trust either government or business except in terms of responsiveness and result* (Hertz, 2001, pp. 251–253).

To sum up, whereas traditional politicians lose credits and take resort to forms of symbolic politics – making promises as if they still were in power – or create a situation of organised irresponsibility – shaking off their liability – the civil society responds by organising collective responsibility, thereby filling the gap that these politicians leave behind. Thus, the civil society is expanding and gaining power. Unlike some critics suggest, Beck as well as Hertz are not promoting the end of representative democracy, but they do believe that other forms of collective action and commitment are necessary to defend our public goods and keep politicians, corporations and international institutions on their toes (Beck, 1992; Hertz, 2001).

Plurality and conflict among adversaries

The neo-republican ideal of a civil society presupposes a particular way of life: an open, active and inspiring culture of debate and struggle for the common good and a common future. As such, political life does not imply a self-securing way of living together with like-minded people. On the contrary, that which constitutes the public domain is the ongoing confrontation with people of different beliefs, opinions and ideological backgrounds. Here one's conception of the good should not be shaped solely within the terms of the established norms of one's own community but in confrontation with those who dissent. Thus, the civil society is profoundly marked by pluralism. Our common commitment to public concerns does not follow from a common core of values or principles but emerges from political practice itself. More specifically, a common practice requires an ongoing 'understanding in agreement'. This agreement is ultimately fluid and can never be made explicit, or exhaustive, because political practice itself is marked by pluralism. Therefore, paradoxically, one could say that it is plurality and conflict rather than moral similarity or agreement that unites us as citizens. Thus civil society theory holds that in public discourse we implicitly agree to disagree, in contrast to liberal theory that renders this agreement explicit in a more or less fixed contract.

Liberalism and communitarianism are largely insensitive to considerations that follow from the social condition of plurality and conflict, dominant in the civil society and at the heart of political practice. In fact, both Rawls and de-Shalit postulate a moral agreement that is assumed to precede political debate, defined respectively as the contractual consensus of pre-political subjects or the moral similarity between members of a pre-political community respectively. From this postulated agreement they derive *prima facie* rules that restrict political action in advance. In her book *The Democratic Paradox* (2000) Chantal Mouffe sharply reveals how Rawls' theory, by postulating an overlapping consensus on which all 'reasonable persons' are assumed to agree, in fact aims at clearing the public space from conflict, struggle and antagonism:

> *Rawls seems to believe that whereas rational agreement among comprehensive moral religious and philosophical doctrines is impossible, in the political domain such an agreement can be reached. Once the controversial doctrines have been relegated to the sphere of the private, it is possible, in his view, to establish in the public sphere a type of consensus grounded on Reason. This is a consensus that it would be illegitimate to*

> *put into question once it has been reached, and the only possibility of destabilization would be an attack from the outside by the 'unreasonable' forces. This implies that when a well-ordered society has been achieved, those who take part in the overlapping consensus should have no right to question the existing arrangements, since they embody the principles of justice. If somebody does not comply, it must be due to 'irrationality' or 'unreasonableness'. At this point, the picture of the Rawlsian well-ordered society begins to emerge more clearly and it looks very much like a dangerous utopia of reconciliation. To be sure, Rawls recognizes that a full overlapping consensus might never be achieved but at best approximated. It is more likely, he says, that the focus of an overlapping consensus will be a class of liberal conceptions acting as political rivals. Nevertheless, he urges us to strive for a well-ordered society where, given that there is no more conflict between political and economic interests, this rivalry has been overcome (...) The way he envisages the nature of the overlapping consensus clearly indicates that, for Rawls, a well-ordered society is a society from which politics has been eliminated* (Mouffe, 2000, pp. 28–29).

Rawls' conflation of consensus and reason precludes the possibility of legitimate dissent in the public sphere, since to disagree with the consensus is to be 'unreasonable'. This argument suffers from extreme circularity: political liberalism provides a consensus among reasonable persons, who, *by definition*, are persons who accept the principles of political liberalism. Herewith, Rawls takes the sting out of public debate, and therefore Mouffe accuses him of 'politicide'[13]: eliminating politics of our lives. At this point an interesting parallel can be drawn with the ideal of sustainable development, which is conceived of, in similar terms, as an antecedently defined consensus on which all 'reasonable' parties should be willing agree. Paul Treanor, a critic of the paradigm of sustainable development, complains that it is hard to have a decent discussion on sustainable development with advocates of this paradigm because those who support sustainable development hardly recognise the possibility of objection. As soon as critics argue that they oppose the ideal of a sustainable development, the advocate will respond that no one is against sustainable development, but that 'we just disagree on its definition'[14]. This anecdote provides a clear-cut illustration of how the liberal focus on consensus tends to de-politicise public debate. To present the institutions and principles of liberal democracy as the necessary outcome of a pure deliberative rationality 'is to reify them and make them impossible to transform' (Mouffe, 2000, p. 32).

Unlike liberals with their focus on consensus, and unlike communitarians who assume a moral similarity between community members, neo-republicans and civil society theorists argue that the conditions of plurality and conflict constitute political discourse. Plurality is not a sign of democratic weakness or lack of community coherence nor a situation to overcome. On the contrary, plurality is the condition that necessitates and inspires us to act and judge politically. Likewise, conflict is not a threat to political discourse that has to be eliminated. Conflict is an element without which no political discussion among adversaries is possible or meaningful. According to Mouffe a democratic struggle among adversaries is of essential importance, if we do not want our conflicts to end in 'a war between enemies'. To illuminate her position on this particular point, she contrasts her idea of plurality with that of Rawls who speaks of 'the fact of pluralism'. By stating this 'fact' as an empirical background for his theory of justice, Rawls refers to a society in which there is no agreement on the ultimate truths, but a diversity of conceptions of the good, a diversity that has to subtended by a contract. However, Mouffe argues, plurality is more than an empirical observation of diversity. Above all, plurality has a symbolic and metaphysical meaning. Plurality marks the irreducible otherness of 'the other': the elusive other who is necessarily excluded by any social arrangement, and therefore forms its constitution. To put it in more practical terms: Mouffe does not believe in the possibility of a power-free dialogue or consensus, because consensus without exclusion is impossible. That is, any consensus, agreement or other social objectivity exists merely by the grace of perspectives that are excluded beforehand. Any social objectivity is constituted through acts of power and exclusion. What is more, as Mouffe suggests: 'Any social objectivity shows the traces of the acts of exclusion which govern its constitution: what, following Derrida, can be referred to as its "constitutive outside" ' (cf. Lyotard, 1984; Smeyers and Masschelein, 2000a and b; Mouffe, 2000, p. 21).

The borderlines between 'we' and 'the other', between 'the inside' and 'the outside' permeate all contracts, commitments, personal opinions and social practices. In this sense, power is omnipresent and inescapable. Mouffe wants us to think of power not as an 'external relation taking place between two pre-constituted identities, but rather as constituting the identities themselves' (Mouffe, 2000, p. 99). Because power is a constituent part of our identity as political actors, we should not strive to ban power from the public sphere, but we should strive to canalise power in such a way, that the dichotomies it produces – we/them, inside/outside, powerful/powerless – are made compatible with our democratic ideals. In other words, we must aim to transform antagonism – war between enemies – in agonism – struggle between adversaries. This transformation cannot be realised merely by designing and monitoring procedural arrangements, since the legitimacy of these arrangements are themselves constituted by exclusion[15]. Therefore Mouffe argues, we have to leave behind the ideal of transparency of power and harmony. Instead, we should acknowledge and actualise within our acts and judgements the productive coalition between power and legitimacy. Thus, justice is not understood in contractual terms, but as a public virtue. Or more specifically, justice comprises a sensitivity towards otherness. Underlying any political judgement and action there should be a personal striving to give voice to those who are excluded by our present judgements and actions. The impossibility of attaining full justice constitutes a continuous incentive to meet the 'other'. This means that a social agreement can never be justified *a priori*, but that its legitimacy has to be gained in every new situation again and again. In a sense, political judgement and action are without precedence. Every political act and judgement should be self-reflexive, i.e. every political act and judgement should produce its own legitimacy. For this reason, we should never adopt a consensus that precedes political debate, because any consensus fixes particular dichotomies, and thereby, limits its legitimacy. Rather, pluralism and conflict should be given a clear field in the public arena to meet the condition of justice.

Although Mouffe does not mention civil society as such, civil society comprises the public space *par excellence* that gives room to conflict and plurality. Civil society theorists eschew all antecedently defined principles that could possibly legitimate an *a priori* limitation of the public space. Most of them will agree that transcendental grounds of justification have lost their credibility in politics. Political arguments can no longer be justified on grounds, principles or authorities that precede or go beyond the common practices of the civil society we live in today. In this sense, the notion of a civil society is an expression of radical secularisation: '(...) the public sphere is an association constituted by nothing outside the common action we carry out in it: coming to a common mind where possible through the exchange of ideas. Its existence as an association is just our acting together in this way. This common action is not made possible by a framework that needs to be established in some action-transcendent dimension: either by an act of God or in a Great Chain, or by a law coming down to us since time out of mind. This is what makes it radically secular' (Taylor, 1995, p. 267). Although political practice is conceived of in secular terms, this secularism does not imply nihilism, since, as we will see in that which follows, it is precisely collective action that shapes meaningful communities.

Since political action is without precedent we cannot rely on it for our justification. Political action has to bring along and express its own justification. Justification has to be realised in the action itself. That means that issues are never 'political' from the outset, because there is no pre-established political agenda of legitimate issues. Citizens and associations of citizens have to *make* things political. Political action places issues on the political agenda. In the eighteenth century, civil society emerged as an independent though secondary counterpart of the nation state: a public, but not politically structured space, where debate took place and public opinions were formed, and wherein the authority of the state was given a mandate. That is, the expanding power of nation states was balanced by the emergence of a critical public to which it should give an account of its actions. Without the mandate of this society the state could not possibly exercise its authority (cf. Taylor, 1995, p. x). What we witness at present, at the outset of the twentyfirst century, is that civil society emerges from the shadow of the nation state and changes the very nature of political action. Where this long term development will lead is uncertain. Whether civil society can ever rely on complete self-regulation without the 'supporting' political framework of a nation state remains to be seen. But neo-republicans and civil society theorists share a common ambition to strengthen this public sphere, and they are looking for ways to change political structures in order to realise this.

But if neo-republicans are not ready to define democratic meta-principles, one might ask, how are they to guarantee democracy? To answer this question, it is important to consider the paradigmatic problem they address. While liberals ask themselves what a state could possibly demand of its citizens, and *vice versa*, what its citizens can demand of the state, communitarian critics are mainly concerned with the problem of how to assure the continued existence of their moral community. Neo-republican theorists, on the contrary, are interested in the question how citizens act collectively in response to collective problems and thereby shape their lives: how they appeal to one another's responsibility by virtue of their being citizens. In this view, a strong sense of collective responsibility arises from horizontal relationships between citizens, rather than from the vertical relationships of citizens to state- or community authorities. Thus, neo-republicans arrive at democratic 'principles' as well, but those will never be given a meta- or *a priori* status, because they cannot serve as a ground of justification. They serve as normative anthropology: what does it mean to be a citizen? As such, the difference between liberal theory and neo-republican theory can be understood as a difference of perspective on the nature of political philosophy. Philosophy should not justify existing political frameworks, but create openings for new ones. Philosophy is not about fixing frameworks but about putting them in motion.

Political judgement

In light of the analysis of civil society and political action in terms of public opinion, pluralism, otherness, conflict, debate, commitment and horizontality just given a neo-republican account of citizenship can be hazarded. As the outline proceeds, the question of future responsibility will be dealt with in particular. In line with neo-republican writings on citizenship, civil responsibility is marked by a particular kind of judgement and commitment (Oldfield, 1990; WRR, 1994; Pettit, 1997; Nauta, 2000; cf. Honohan, 2002). As a form of practical wisdom, political judgement allows people to distance themselves from their roles in society, reflect on the structural factors that constitute these roles, and determine their course of action from there. In case of environmental concerns, for instance, there is a need to reflect on our role as consumers and the ways in which consumer markets direct our 'free choices' and behaviour in directions of which we might not approve. So, in public discourse, we need to make judgements about these issues, at a certain distance from our roles as consumers. This emphasis on the public nature of judging is in line with Arendt's statement that moral judgements are always expressed in front of a – real or imagined – audience of others whose recognition one seeks. Our judgements would be meaningless unless there were others present to recognise them. Consequently, to withdraw from moral judgement is tantamount to ceasing to interact, to talk and act in the human community (Arendt, 1982).

However, this audience does not speak with one voice, but shows a concert of voices. Within a civil society, Arendt argues, citizens judge in light of plurality: 'Action, the only activity that goes on directly between men without the intermediary of things or matter, corresponds to the human condition of plurality, to the fact that men, not Man, live on the earth and inhabit the world (…). Plurality is the condition of human action because we are all the same, that is human, in such a way that nobody is ever the same as anyone else who ever lived, lives, or will live'

(Arendt, 1958b, pp. 7–8). In public discourse the differences between human beings flourish and become productive. That is, our expressions of difference – our words and actions – acquire meaning and influence by means of their immersion in an intersubjective web of meanings and relationships; a common world or language. Unfortunately, as a consequence of this 'immersion' in a complex interplay of forces the agent loses control over the meaning of his words and actions. The consequences of our actions proceed in directions we can not possibly predict. As such, unintended (side) effects have an open field. This recognition of the boundlessness of human action and power leads Arendt to define unpredictability as a fundamental frustration of political action. Our actions engender a chain of reactions in a direction that is beyond our grip. The effects of this particular uncertainty are reinforced by another frustration of political action, the frustration of irreversibility: that which is done within the public arena cannot be undone. That means that we are not able to nullify the unintended effects of our actions. These two fundamental frustrations of unpredictability and irreversibility are in Arendt's view inherent to the narrative character of action. Only in the told story that is reconstructed by hindsight, do the acts of the protagonist acquire full meaning. As long as the story proceeds, the plot is by definition unknown. (Arendt, 1958b, pp. 236–247).

These two frustrations turn political judgment making into a risky game, somewhat like gambling. How are we to judge responsibly, if we do not know what the consequences of our actions will be and are unable to correct our actions if these consequences turn out to be horrible? Throughout western political history many remedies have been tried to eliminate these weaknesses but most of them consisted of eliminating plurality. This was what Plato did when he proposed to render all political power to a king philosopher. And according to Arendt, this is what those philosophers do who turn the *praxis* of politics into a *poeisis* – an activity of 'making' or 'managing' social affairs. Arendt argues that the only acceptable remedies are to be found in the political virtues of promise and forgiveness. To overcome the weakness of unpredictability, people make promises and lay them down in laws, treaties and agreements. By articulating promises in public documents, we hold one another responsible for the continuity of our intentions over time. And the human power to forgive should balance the irreversibility of political action. Without this power to forgive we would probably feel paralysed beforehand by the prospect of forever being chained to the unintended effects of our past actions and judgements. In my view Arendt marks the contours of an intergenerational ethic by posing a specific relationship between the faculties of promise and forgiveness: 'The two faculties belong together insofar as one of them, forgiving, serves to undo the deeds of the past, whose "sins" hang like Damocles' sword over every new generation; and the other, binding oneself through promises, serves to set up in the ocean of uncertainty, which the future is by definition, islands of security without which not even continuity, let along durability of any kind, would be possible in the relationships between men'. (Arendt, 1958b, p. 237).

Arendt would be the first to admit that these remedies are hardly airtight and perhaps even somewhat naïve. However, if we adhere to plurality and political autonomy as democratic values, we have to make do with the powers we have. But if the basic pretensions to predictability and controllability are to be given up, the prospects of judgement in general and those of future responsibility in particular look worse than ever. How can we hold one another responsible for the consequences of our actions if they are not in our hands? Obviously, the conditions of political action force us to understand civil responsibility beyond the instrumental terms of accountability: being held responsible for the intended consequences of one's actions in light of pre-given aims and standards. In her *Lectures on Kant's Political Philosophy* (1982) Arendt shows that political judgement comes into play when the standards or aims which constitute the grounds on which one ought to judge, are not self-evident or fixed beforehand; when conflicting claims and stories about reality have to be weighed against each other. This weighing cannot be resolved on the grounds of any pre-given standard, exactly because such a standard is not available to us. Political judgement, like aesthetic judgement, is without precedent. Each particular case is unique in the sense that it calls for its own standards, just as a work of art thrusts its own standards on us. In this sense political judgement floats by 'thinking without banister', as Arendt argues.

To acknowledge the plural nature of political judgement is to say that we cannot rely on a pre-given framework of shared values underlying our current debates. In our attempt to convince others of our point of view, we cannot appeal to a common good or common conception of justice, preceding or transcending the arguments we express here and now. To judge prudently then, is to deal with a plurality of opinions, and to deal with the subsequent conflicts of loyalty. In order to fathom the precise nature of political judgement, Arendt reverts to Kant's political philosophy. However, she does not elaborate on Kant's ethics of *The Critique of Practical Reason* (1788), which would be the most evident point of departure, but on his ideas on aesthetic judgement as developed in *The Critique of Aesthetic Judgement* (1790). Arendt was one of the first thinkers to recognise that political judgement can be understood analogous to esthetical judgement. Rather than an activity that consists of applying principles or following rules, making judgements in esthetical and political matters is an art of 'enlarged thinking' (*eine erweiterte Denkungsart*):

> *The power of judgement rests on a potential agreement with others, and the thinking process which is active in judging something is not, like the thought process of pure reasoning, a dialogue between me and myself, but finds itself always and primarily, even if I am quite alone in making up my mind, in an anticipated communication with others with whom I know I must finally come to some agreement. And this enlarged way of thinking, which as judgement knows how to transcend its individual limitations, cannot function in strict isolation or solitude; it needs the presence of others 'in whose place' it must think, whose perspective it must take into consideration, and without whom it never has the opportunity to operate at all* (Arendt, 1961, pp. 220–221).

Judgement is thus conceived of as the ability 'to think in the place of everybody else'. This capacity and willingness to take on multiple perspectives, she insists, is a way of situating and regarding oneself as a political being operating in the public realm. My subjective judgements are only fleeting, subjective impressions until they are thrown into discourse with others. Political judgement for Arendt, like aesthetic judgement for Kant, necessarily implies an attempt to 'woo the consent of everyone else'. Judgements are formed first by reviewing my conceptions through others' perspectives in the hope of arriving at a mutual agreement about what will appear in our world. As such, judgmental thinking comprises a form of representative thinking; our imagination and reflection helps us to present the absent, or in other words, to represent the perspectives of those who are not present. Judging then consists of viewing a particular case from different perspectives, by taking the perspectives of as many as others as possible (Arendt, 1982).

For Kant a particular judgement cannot be subsumed under a general rule or predicate, but the judgement itself reveals a general predicate that we will never be able to articulate exhaustively. In line with this insight, Arendt points at the *exemplary* validity and power of judgements in political discourse. Judgements that appeal to particular events sometimes reveal a strong general idea that cannot be expressed otherwise. Particular events in recent history have fuelled public debate on environmental issues and shaped our awareness of environmental problems in an unprecedented way (Arendt, 1982, thirteenth lecture). If, for instance, we take into account the warnings of the Club of Rome, the sinking of the *Rainbow Warrior*, the nuclear disaster in Tsjernobyl, the conflict about the Brent Spar, the shipwreck of the *Exxon Valdez*, the outbreak of foot and mouth disease, the revolt of the Zapatista's against the NAFTA treaty, or the recent protests in Seattle and Genoa, then we must acknowledge that historical events like these have contributed to our environmental awareness and ability to act in a way that a principle or universal ideal would never achieve. As such, these events are manifestations of a non-universalisable freedom.

Within our faculty of judgement social imagination plays a vital role: as we anticipate the possible judgements of imagined others we tune in to the *sensus communis*. Literally speaking, the *sensus communis* refers to the common sense we share as members of a political community, or more precisely, the sense that creates a community. This is not a matter of formal agreement or a shared stock of substantive values but a common way of responding to things that appear to us. A *sensus communis* does not necessarily imply that we act in the same way – although this will often be the case – but that we experience reality in a way that serves "to direct us in the common affairs of life, where our reasoning faculty would leave us in the dark" (Shaftesbury, cited in: Gadamer, 1960, p. 31). As such it is not a common way of doing things but a common way of experiencing the things we do (Gadamer, 1960, pp. 24–35; Arendt, 1981, pp. 50–53; Arendt, 1982, pp. 105–106). In the particular meaning Arendt attributes to the *sensus communi*s, it functions as a warrant for impartiality in cases of political judgements. By taking the perspectives of others, that is, by 'enlarging' our point of view in such a way as to include the possible standpoints of fellow citizens, we transcend the subjectivity of our personal judgement and open up the possibility of a general standpoint. In this specific sense esthetical judgement is characterised by disinterestedness. Thus understood, impartiality is not typified by the perspective of the arbiter standing above the parties, but rather by the perspective of the dedicated parent who strives to adopt a general standpoint in cases of conflict. The more perspectives they take into account, the broader their scope of judgement and the more general or 'impartial' their standpoint[16]. Again, taking responsibility for one's judgements is conceived of in terms of openness to a plurality of voices. Rather than a matter of justifying our judgements and actions in terms of the standards on which they rely or the aims to which they are supposed to contribute, taking responsibility is like the kind of responsiveness we exhibit in profound story-telling. In order to make our judgements and actions sensible to ourselves and others, we tell stories in which our articulation of the actions and motives of the 'I' solicit the recognition of a largest possible audience (Arendt 1982; cf. Altieri, 1994; Van Nieuwkerk and Van der Hoek, 1996; Whiteside, 1998, p. 36).

Judging the status of future generations

If the *sensus communis* constitutes a common sense we share as members of a civil society, who are the 'we' on behalf of which we judge and speak? Are future generations included in this community? Obviously, there is no pre-existent community on the rules or standards of which the legitimacy of our judgements rely – like liberals and communitarians assume – but it is the other way around: deliberative judgements shape communities around issues of a common concern (cf. Whiteside, 1998, pp. 35–38). This means that there are no fixed or fundamental limits to our political community beforehand, for every 'other' who is not yet included in the 'we' might be a potential partner in conversation and *ipso facto*, a potential other whose perspective I should take into account in my search for a general standpoint. Unfortunately, there is at least one practical limit to the scope of our political community and that is a temporal limit. 'We' comprise a community with a common sense by virtue of the fact that we are potential partners in conversation. This implies that 'we' should at least be contemporaries. The others whose potential perspectives I ought to take into account in my search for a general standpoint are *real* others, with whom I should be able to experience a certain reciprocity. To be more precise, what is lacking in my relationship to future people is not a reciprocal relationship in terms of strict mutuality like Rawls assumes or a common good as de-Shalit argues. Moreover, the particular kind of reciprocity Arendt implies is operative in our engagement in a common discourse or practice.

In my view, this reciprocity consists of a reciprocal quest for recognition as discourse members. For, if we understand judgement in terms of anticipated communication, like Arendt does, then we have to recognise one another as potential communication partners in order to woo each others' consent. Unfortunately, future generations are beyond the scope of such a potential dialogue. Right here and right now, a dialogue with people who will live in the distant future is inconceivable and beyond imagination in the practical sense Arendt has in mind. We cannot anticipate the potential judgements of future descendants in a way that constitutes our present judgement, because, as we have seen in the foregoing analysis, one's contemporary political community is the ultimate bedrock on which all our political judgements and actions are based. Arendt's practically empirical understanding of the *sensus communis* – intimately tied to our sensual experience of participating in public discourse – does not allow for an abstraction from this community towards a potential future community. For this reason future generations are excluded from our political community by neo-republicans as well.

This conclusion might appear to be a drastic disillusionment for those who expected neo-republican theory to provide a clear-cut alternative for the fruitless attempts of both liberal and communitarian theory to include future generations. One might wonder why we have taken this route if it does not seem to lead us anywhere. Well, as I will argue, our study of neo-republican ideas until now has indicated why every attempt to include future generations as fellow citizens is doomed to fail. A closer study of the necessary limit of political judgement and responsibility will disclose an alternative understanding of future responsibility.

What has become clear in the foregoing analysis is that every attempt to include the rights, needs, goods or potential judgements of future citizens throws us back upon our own devices. In our present judgement on what will count as a liveable world for future generations and our estimation of what their needs will be, we necessarily rely on what we sense as 'liveable', 'inalienable' or 'dignified' within our current political community and horizon of sensibilities. We reflect on the possible implications of our present behaviour for people in the future on the grounds of the deliberative judgements we share with contemporaries with whom we act together. Our future imagination and expectations take their cue from our *sensus communis*, a common sense that is tied to a particular time and place. Thus, our diagnosis corresponds closely to that of communitarians like de-Shalit. However, unlike de-Shalit, we do not assume that posterity will be like us, and that they will share our conception of the good life. We are not connected with future citizens by means of a common good. Moreover, we are connected only loosely by means of evolving practices that constitute the evolution of our political community. As historically situated beings, we are connected with citizens of the nineteenth century in the sense that the public practices in which we participate bear the traces, for instance, of their efforts to realise universal suffrage. Just as they helped to shape the current practices in which we participate, so our participation will help to shape the practices in which the citizens of the twenty-third century will participate. But to say that we are connected with future citizens by means of historically evolving practices is not the same as recognising them as participants in the same practices.

This insight allows us to indicate even more precisely why the activity of 'enlarged thinking' (inherent in political judgement) can only be 'enlarged' towards persons existing at present. Arendt writes: 'this enlarged way of thinking needs the *presence* of others "in whose place" it must think, whose perspective it must take into consideration' (Arendt, 1961, p. 220; italic mine). In my view, the others' *presence* is required, because my search for an impartial standpoint is motivated by my desire for their recognition. We anticipate communication and take multiple perspectives because we desire recognition as participants in public discourse. And this desire for reciprocal recognition emerges in common practice. It is the realistic possibility of gaining recognition from discourse members that motivates our striving for generality and impartiality. Thus, the contemporary political community *casu quo* sensus communis forms the warrant for impartiality. Of course, we might desire recognition from those who are not yet included in public discourse. After all, the boundaries of our political community and sensus communis are not fixed. But we can be sure that those who do not yet exist cannot answer or mirror our desire for recognition. And in the absence of any answer, a lasting desire for their recognition as discourse members is not likely to emerge. Therefore, anticipated communication with future generations will remain illusive and cannot be a source for political judgements.

Within public discourse, appeals to the needs, rights or goods of future generations ought to be answered with suspicion. Those who pretend to speak on a behalf of future generations often hide behind future generations in order to pursue their own projects. As a particular rhetorical way of phrasing arguments 'the interest of future generations' lends itself too easily for conflation with our own personal preferences and ambitions. On the internet, for instance, we find strong proponents as well as opponents of an eugenic policy designed to decrease the world population and increase our evolutionary fitness. Both of them refer to 'the interests of future generations' as though these arguments would sweep away any counter-argument. Proponents of an eugenic policy argue that future generations are entitled to a certain 'genetic health', whereas opponents argue that they have the right to live in a 'natural world', free of manipulation and genetic doctoring. What we witness here, is that future generations have come to function as a projection screen for our collective fears and hopes. However, those fears and hopes themselves remain unarticulated in public discourse because the appeal to future generations is regarded as self-evident.

Thus, instead of judging properly by taking multiple perspectives of fellow citizens and borrowing arguments from the intersubjective *sensus communis* underlying our political community, these participants take refuge behind an external authority: the chain of future generations. And since no contemporary citizen is in the position to trace the origins and authority of these arguments, the strong appeal to future generations is similar to an appeal to God: these arguments have a metaphysical bias, in the sense that their legitimacy depends on the invocation of a transcendent authority external to human discourse. In fact, the idea of sustainable development seems to imply a secularised form of the biblical stewardship: instead of God calling on our responsibility as stewards of our world, it is the infinite chain of human generations that gives the world on loan under our trust. The difference between the two is a difference of degree rather than a difference of kind. We do not have to go as far as deliberative democracy theorists, who state that these arguments are not at home within political discourse and should therefore be denied status (cf. Korthals, 1994, p. 51; cf. WRR, 1994). However, we should acknowledge that these arguments either have no chance to solicit agreement among discourse members, or tend to mystify the issues at hand and de-politicise public debate. So, as a source of political judgement future generations are suspect.

In pursuit of continuity of our civil practices

The previous analysis indicates that our relationship with future generations cannot serve as an authentic *source* for judgement, but leaves open the possibility that their claims are an *object* of judgement. That is, we cannot include future people in the 'anticipated communication' inherent in political judgement as anticipated communication partners, but their claims could be among our subjects for discussion. For instance, in our current debates on what energy policy to choose we would presumably consider the interests of future generations without meeting them as fellow citizens whose actual consent or participation in deliberation is required. Rather than fellow citizens, future generations should be regarded as those who will be affected by our present judgements: the beneficiaries or the parties injured. However, as I have argued before, most public appeals to the needs, rights or goods of future generations are of an exclusive rhetorical nature. Nevertheless, these appeals are highly attractive in public discourse. They make a strong demand on us. Therefore, we cannot do away with all these claims as merely rhetorical window dressing, but we ought instead to pay closer examination to what is implied in our recognition of any claims made on behalf of posterity.

While we recognise claims on behalf of future generations as making a demand on us, we do not anticipate communication with them. We may perhaps wonder how our potential grandchildren or great grandchildren might react to our present judgement. But as soon as we leave this inner circle of family and friends, there will be no particular others who make an immediate demand on us like contemporaries do. Moreover, the 'call' we experience emerges from our practical involvement in those public practices, projects, movements and traditions to which we are committed at present. Our personal investment in these practices and our dedication to its purposes imply an open time horizon. For instance: citizens participate in environmental pressure groups because they believe that their influence on decision-making processes can make a difference. Their involvement is meaningful in light of the belief that the particular future of these processes is not fixed but open-ended, and at least partly dependent on our present judgement and action. What is more, the participants will experience their collective action as meaningful, even apart form the effects in the future. As a worthwhile practice they will care for the continuation of this practice in the future; as such, their commitment implies a desire that this commitment be sustained or carried over to future people. To examine this claim, Visser 't Hooft wonders what would happen to our practical commitment if we were to realise that we were living in the final days of mankind. He suspects that an immediate feeling of absurdity would take possession of us and our activities. Our participation in these practices would become meaningless, he concludes, because our commitment is partly premised on trust in the continuation of these practices in the indeterminate future. Inherent in our present commitment is a claim that takes possession of a particular future of indeterminate length. As such, basic trust in a common future is among the very conditions of our existence as citizens (cf. Achterberg, 1989/1990, 1994; Visser 't Hooft quoted in Van Putten, 1999).

I deliberately speak of *continuity* of practices rather than 'sustainability', 'durability' or 'consistency' because the first term leaves more room for unprecedented change and difference. Furthermore, the term 'continuity' seems more adequate to express experiences of the inside practitioners, rather than the qualification of those who observe or study their commitment in the more objective terms of the outsider. Take for instance, the identity change that the environmental movement has undergone over the last thirty years. Environmentalists themselves might experience their present commitment in well-established funding organisations as following on in a process of continuity from their radical involvement in the late seventies, whereas the media and public criticise their action over a period of time as being 'inconsistent' or 'untrustworthy'. The same discrepancy applies to reverse situations in which the public perceives the actions of a particular environmental group as consistent with their image of this group over all these years, while I myself have lost my commitment: their actions are no longer a continuation of my authentic desires anymore. These considerations touch upon considerations of identity-formation, i.e. in order to be ourselves and keep our commitments and 'selves' alive, we have to change (cf. Bransen, 2003). Civil practices are worthy our concern, because they contain within themselves the means to carry over what we regard as important in our modes of behaviour into new situations we face. The insider's perspective is of greater importance for environmental ethics and politics, because it is our personal experience of continuity inherent in our fulfilment that directs our attention to the future and obliges us to care for the future (Winch, 1958).

This basic assumption of a dynamic continuity underlying our involvement in civil practices marks an importance difference with de-Shalit's communitarian approach. Neo-republicans and communitarians both locate the sources of future oriented responsibilities in our present desire for recognition of significant others. However, whereas de-Shalit argues that those 'others' include future people, for us, the others are necessarily contemporaries. But even more important are the implications of the underlying contrast between the communitarian insistence on moral similarity and the neo-republican stress on plurality as hallmarks of a political community. An extrapolation of the neo-republican assumptions teaches that what we share with contemporaries as well as future generations is not a common good or morality. Instead, we are connected by means of an evolving *sensus communis* or form of life that consists of multiple practices, which differ as well as overlap. These practices change analogously to Wittgenstein's tables of family resemblances (Wittgenstein, 2002, p. 111). Acknowledging the intrinsic quality of difference and change in our judgements leads us to recognise that we do not necessarily solicit agreement with others, but we consider their possible standpoints and judgements as we go along. Seeking judgement, I try out my views and thereby test their strengths and limits. This means that we might sometimes head for conflict rather than for consensus. Not (only) because we want to compete with others or because we have conflicting interests, but just like communitarians would argue, because we strive for a genuine continuation of our 'selves' and *ipso facto*, the continuation of the civil practices that constitute our selves.

Future imagination and immortality

A remarkable similarity between the liberal and neo-republican view on future responsibility is that both views regard future generations as possible beings with hypothetical needs rather than kindred heirs with familiar preferences. To us, contemporaries, future generations remain faceless and nameless. The only familiarity between future and contemporary generations, and the only relationship between us, is created by our ability to hypothesise. However, liberals and neo-republicans would not agree on the particular nature of our hypothethical thinking. According to Rawls, we should calculate what future generations are entitled to by means of a rational reconstruction of those necessary needs whose fulfilment would be conditional on leading a meaningful life – independent of time and place. Neo-republicans, on the contrary, do not believe in the possibility of such an a-historical concept of personhood, and would probably give more weight to the act of imagination. As Arendt argues, in line with Kant's aesthetics, our faculty of judgement consists of a twofold mental operation. To begin with, it is through imagination that that we can present to our inner sense that which is absent: the object of judgement. Our imaginative powers enable us to picture possible future worlds, in such a way that we are immediately seized by its beauty or ugliness, its goodness or badness, its truth or illusive nature. Those sensations are highly subjective and can hardly be represented in language. Therefore, subsequently, we perform an operation of reflection on that which has previously been presented by our imagination. As indicated earlier, this reflection comprises an act of taking multiple perspectives and making our reflection communicable to others by anticipating on a common sense we share as a community: a *sensus communis*. Thus, imagination as well as reflection plays a vital role in the transformation from subjective taste to judgement by moving back and forth from the inner sense to the common sense (Arendt, 1982).

Our visions of the future will always be rather vague and rarely well-defined. They cannot be designed out of a detached position of contemplative isolation, as Arendt makes clear, because our imagination requires a common sense in order to be stable and communicable. Thus, our imagination of the future will emerge from our involvement in every day activities. We make them up as we act and go along. Sometimes they are alluded to or articulated in poems, paintings or novels, but generally, they remain unsaid. As such, these visions remain elusive. Perhaps it is their elusiveness that lends these visions their attractiveness. We keep on chasing them but as soon as we try to pin them down, slip through our fingers. Within the political realm, our imagination is generally triggered by desires and fears that adhere to our personal involvement in public practices. Imagine, for instance, a technically skilled citizen who participates in a campaign for hydrogen energy applications in public transport. This involvement presumably brings with it a strong desire to change our way of travelling and our systems of public transport in an unprecedented way. But as that person's desire grows stronger, so too will the fear that the mission will fail. The activity of imagining that takes its cue from this experience consists – at least partly – of an extrapolation of particular rules underlying our current practices leading into an intuitive, though unknown direction. Suppose, for example that the advocate of hydrogen energy enjoys strong support for his articles on this subject in local newspapers and he gets invited by local energy companies, then the imagination concerning his future efforts will probably move from the local newspapers to the national newspapers to international publications. This is a loose extrapolation of the rule underlying his past experience – gaining public support and pressure by publishing articles – into the indefinite future. Of course it might work in the other direction as well. Suppose, for instance that his articles are repeatedly refused for publication, then he will probably extrapolate the rule underlying this disillusionment into the future: he might anticipate resistance and lack of interest in public discourse. Obviously, the extrapolation that is part of our imagination takes place within human practice and discourse, because only if qualities are validated through public deliberation, does it become possible to apply them in new ways. Thus, the focus of our imagination will be on future practices rather than future people. Future people remain imagined strangers to us. They inhabit a world we dream of, and we sometimes allude to this possible world but we cannot derive any obligations or responsibilities from these worlds without losing touch with present reality. What they can do, is make a demand on us to consider what in our life is worth sustaining.

There is significant degree of imagination inherent to political action, thus neo-republicans argue. In *The Human Condition* (1958b) Hannah Arendt argues that in the ancient *polis*, political action was thought to be motivated by a profound striving for immortality. For the ancient Greeks and Romans the only way to overcome our mortality would be to act gloriously on the public stage and to gain honour in the eyes of one's fellows and heirs. Arendt recalls the idea that, in the ancient meaning of the word, 'dying' was regarded equivalent to 'ceasing to be among men' (Arendt, 1958b, p. 20). To be immortal in this sense would be to stay alive in the memory of others after one's death. The idea that the public sphere can provide a kind of 'earthly immortality' in the face of human mortality dissolves, however, with the discovery of the fragility of the public sphere itself (Arendt, 1958b, pp. 28–32, 64). What might be left of this ancient idea in post-modern times is our desire to be remembered by future generations in a particular way. We need a sense of continuity – a sense that our worries, hopes and dreams have not been in vain because earthly life goes on, and our worldly activities and projects will be carried on. In short, the idea of a virtual bond with the past and future world seems to be an anthropological *sine qua non* for humans to make sense of their lives. We want to leave behind a world, which we perceive as worthy our remembrance.

The meaningfulness of these considerations reveals that there must be some other kind of reciprocity between us and those unknown people in the future, who will live centuries from now, even though this reciprocity in is not in any strict sense, procedural. Within the public sphere, citizens want to leave something behind that corresponds to their own understanding of self. Though even this desire might sound somewhat megalomaniac, namely the idea that we, contemporaries desire recognition from future generations is not completely beyond the pales of imagination. To give this intuition a less narcissistic slant, it might be helpful to explore the distinction between 'grand' and 'small immortality', as pictured by Milan Kundera in his novel *Immortality* (1990). Both aspirations imply a refusal to die in the present and with its presentday cares and a wish to transcend oneself. However, the aspiration to grand immortality consists of a desire to exhibit oneself on the stage of human history and rank oneself among the 'big men' in some Hall of Fame, while the aspiration to small immortality consists of the modest desire to be remembered by those who have known us during our lives (Kundera, 1990). We strive for continuity of our relationships and practices because we want to maintain our 'self', perhaps even beyond our death. Conceived of in these terms, our inherent striving for immortality might imply a particular caring attitude: care for the continuity of our practices and the sustainability of our world. In chapter four we will examine the status, possibilities and limits of this idea more closely in connection with environmental responsibility.

In the neo-republican view on future responsibility sketched in this chapter, the main focus shifts from the rights and needs of future people to our present worries, cares and desires on the grounds that we imagine a common future. As we consider the implications of our current practices for future generations, we will imagine a particular future world, but rather than calculating what would be in the interest of its inhabitants – future people – we have to judge for ourselves: what kind of world would we want them to live in? Which practices are worth carrying over? Which world would be worthy our remembrance? By taking responsibility for the world we want to leave behind, and by safeguarding the continuity of our worthwhile practices, we do not need to hide behind hypothetical rights of future people. Judged from a neo-republican view, that would be a means to run away from our responsibilities. In this approach toward future responsibility, future generations are only of indirect signficance. When we say that we anticipate their possible standpoint, our judgement is in fact a function of our present commitments. Underlying this judgement is our striving for continuity of what we care for. The strong conviction that future people are entitled to a dignified existence similar to ours is derivative of this primary judgement.

Why survive?

Our hypothetical relationship to future people in itself cannot be an ethical source to act on their behalf. However, imagining their future existence calls on our present feeling of responsibility. To be more precise, their image forces us to make up our minds about what we want to leave behind. Simultaneously, their imagined existence prevents us from doing things that might reduce their life chances or ruin their world. Now, the core question is: what is the status of our conviction that we ought to strive for continuity of human practices? What is the metaphysical status of the ideal of sustainability (broadly conceived of in terms of continuity of practices)? In the end, our answer to this question is dependent on our answer to the question of human survival: why should there be a future world, why should there be future generations, why should there be future life at all? Is there a duty to survive? This question is, at least partly, of a metaphysical nature, and as such, philosophers have touched upon this question.

But first of all it is important to stress the logical point that, in a particular sense, the future is always given in our present horizon of action, as Winch makes clear in his analysis of meaningful behaviour. To say that behaviour is meaningful is to say that this behaviour is intelligible in terms of the modes of behaviour which are familiar in our world, and that it was governed by considerations appropriate to the rules of its context or 'language game'. Inherent to Winch's conception of meaningful behaviour as rule-governed behaviour is the notion of 'being committed by what I do now to doing something else in the future'. Take for instance his example of bookmarking: '(…) if *N* places a slip of paper between the leaves of a book he can be said to be 'using a bookmark' only if he acts with the idea of using the slip to determine where he shall start re-reading. This does not mean that he must necessarily *actually* so use it in the future (though that is the paradigm case); the point is that if he does not, some special explanation will be called for, such as that he forgot, changed his mind, or got tired of the book' (Winch, 1958, p. 50). In other words, it is because I intend to do something tomorrow that my action today is intelligible and meaningful. But while I am committed to doing something in the future by doing something now, my future action is not determined by my present action. Winch's point is that all human action borrows meaning from this commitment to an inescapable horizon of past and future action. However, this leaves the metaphysical background question – should there be a future? – untouched.

Now we can turn to the metaphysical question of what commits us to strive for survival of human life on this planet. To start with a brief analysis of the implications of Kant's philosophy on this particular point: the maxim underlying our intention to spoil the life conditions of future generations would obviously fail the test of universalisation, implicit in the categorical imperative. We cannot want to contribute to the extinction of mankind and at the same time want others to do and have done the same, because this striving would pre-empt any contribution to existence or non-existence whatsoever. Therefore, universalisation of this particular maxim leads to contradictions on a volitional level we should not tolerate. As such, Kant implicitly defends a duty to contribute to human survival. However, for Kant, future obligations do not amount to future *people* or their well-being. Moreover, these obligations follow from the realisation of talents inherent to humanity as a species, guided by the law of progress. As Kant articulated in his *Idea for a Universal History with a Cosmopolitan Purpose*: 'human nature is such that it cannot be indifferent even to the most remote epoch which may eventually affect our species, so long as this epoch can be expected with certainty' (cf. Arendt, 1982, p. 118; Van der Wal, 1989, p. 160; cited from: Li, 1994)[17].

Apel underlines the imperative 'that there be humanity', although he does not share Kant's metaphysical idea of progress but departs from the assumption of human beings as rational beings. For him, our collective duty to survive follows from the transcendental conditions of the communication community of which we are always already a part. Were we to collectively agree to commit suicide, then this agreement would conflict with the duty, presupposed in our argumentation, to anticipate the realisation of an ideal communication community. The arguments leading to this radical decision would go against the very conditions that render these arguments valid. Consequently, in every attempt to deny a duty to survive, our commitment to survival is already implied. So, for Apel as well as Kant, survival is a matter of consistency with a transcendental argument (Apel, 1996). Though an ingenious and interesting point of view, neo-republicans would not go along with Apel, since this transcendental line of thought subsumes the plurality of possible voices in public discourse under a universal principle or rule of consistency and thus limits the range of legitimate thought. In line with Wittgenstein's view on the nature of language games, neo-republicans would stress that the application of a principle or language rule is not implied by the rule or principle itself. The relationship between a rule and its application in language is contingent in the sense that the rule could have been applied otherwise. Rules can only be reconstructed retrospectively. Now, Apel might insist that this arbitrary relationship indeed holds for most rules as part of a language game, but that there is a special class of rules that transcends the diversity of language games by virtue of the fact that these rules make possible this diversity without being part of it – for instance – the ideal or principle of diversity itself. They are rules-*about*-language games rather than rules-*of*-language games. These *transcendental* rules, so Apel claims, can be held universally valid and should be exercised as such. However, it does not make sense to distinguish between rules-*of*-language games and rules-*about*-language games in the way proposed here, because even these more abstract rules-*about*-language games – the rules of consistency and survival that will indeed have a wide appeal – are meaningless unless they are expressed in a particular practice according to the rules-*of*-language games that are operative within this practice. As soon as rules are applied, the practice of application requires guidance by rules-*of*-language games. Thus, transcendental rules can only maintain their *transcendental* status as long as they remain unapplied *casu quo* unspoken, and as such they can only function as an empty category or a fictive entity. This is why language rules can never be conceived of as necessary conditions for a particular practice or language game in advance. Consequently, there would be nothing irrational or inconsistent to our collective decision to 'finish' humanity for, by communicating, we do not commit ourselves to antecedently defined rules of survival underlying our language games (cf. Mouffe, 2000; Wittgenstein, 2001; Smeyers, 2002).

Instead of begging the possibilities of human language and reason, Jonas locates the duty to survive in our very being, that is, in our human being as part of humanity. Inspired by the Aristotelian idea that all living things strive towards the realisation of their own potential in accordance with an inherent purposiveness or *telos*, Jonas argues that we are summoned to respect the *telos* immanent to all life. He elevates the existence of humanity to the level of the highest realisation of nature. This status brings along a special responsibility for mankind; the responsibility of guarding over the flourishing of all life that surrounds us and sustains our own flourishing. The imperative of responsibility does not originate from human or divine will, but from the immanent claim of the intrinsic good to its own realisation. Every being ought to be, because of the realisation of its potential good. Jonas perceives the actual striving of an organism for survival as an articulation of the fact that this being wants itself. Thus, he derives an 'ought' from an 'is' by pointing at the underlying teleological self-affirmation of being. There is an *ought-to-be*, beyond the *is* (Jonas, 1984; cf. O'Neill, 1992).

Recognition of the realisation of such an intrinsic good immanent to earthly life would be very helpful to understand and ground our duty to survive. But, even apart from the apparent anthropocentrism present in this line of thought, there are profound reasons why we should refrain from adopting Aristotelian metaphysics as a ground for the imperative for survival. My main objection could be stated in terms of plurality. In order to accept that all that is, ought to be, one has to embrace the 'metaphysics of presence'. Derrida employs this label to indicate the western intellectual tradition, going back to Parmenides and Plato, that tells us that beyond humanity, and immune to historical and cultural change, there is something to which humanity owes respect. We assume that there is some present world, which we then represent. This is something which we have a duty to make clearly and distinctly present to our minds. Derrida attacks this privileging of presence by arguing that we can only have something that is through difference. Any identity requires some system of differences in order to be an identity, in order to have a certain character, which would remain the same through time. To perceive something as present, or as having being, means that we give it some form above and beyond the flux and the plurality of its appearances. It is, as Derrida has put it, 'a fixed presence beyond the reach of play'. Sometimes this thing is called God, sometimes the intrinsic nature of reality, sometimes moral law and sometimes the structure of human thought. Throughout history, the metaphysics of presence have proven to be dangerous and extremely violent, because all are subsumed under the regime of one 'transcendental signified'. Such a signified cannot be sensitive to a plurality of perspectives. Rather than owing respect to that which is present, we should do justice to that which is presently absent, by means of exclusion: 'the constitutive outside' as Mouffe borrowed from Derrida (Derrida, 1995; Mouffe, 2000).

The previous shows that there is no duty to survive inherent to our practices as a language community or to our being human as such. Perhaps it is meaningful to think of this duty and the ideal of sustainability – urging us to strive for continuation of our worthwhile practices – as an inescapable human response to the perspective of an infinite and open future. For us, this unthinkable future embodies the 'constitutive outside': it enables us to make judgements and it (partly) constitutes these judgements, but, at the same time, it is elusive. Not only the future itself is elusive, but our imagination of the future eludes all human judgement and action. In our judgement and action, we anticipate possible futures, but we will never be able to articulate our future imagination fully. Obviously, our future imagination transcends the reasons we express for having them. In my view, that is because we respond to the 'constitutive outside' in a way that is not transparent to ourselves. It would be an illusion to think that our future expectations, commitments and desires can be surveyed by oneself altogether. To a certain extent, they remain opaque. That is why we sometimes catch ourselves longing for something or caring for something about which we were not previously aware. Many philosophers have wondered why we tend to be most convinced of the value of something when the existence of the object of our care is being threatened: 'We become aware of the importance of the ozone layer only when the pursuit of individual goods begins to destroy it, and of languages when they are dead or dying. We regret the loss of countryside or wilderness only when these have been dramatically reduced by urban and industrial expansion' (Honohan, 2002, p. 159). Hans Jonas generalises this insight by stressing that 'we know the thing at stake only when we know that it is at stake' (Jonas, 1984, p. 27). This insight leads Jonas to acknowledge the 'heuristics of fear'; more precise than positive mental states like joy or desire, fear of specific dangers in the future points our attention precisely to those vulnerable goods we really care for. Perhaps, John Cale was right after all, when he wrote the song *Fear is a Man's Best Friend* (1974)[18].

Whereas fear can be an indicator of what we care for, it appears to be a bad counsellor when it comes to motivating environmental education (this question will be addressed in chapter four). But apart from these educational considerations, the heuristics of fear point our attention to a different way of conceiving our relationship with future generations as future members of our *political* community. As suggested earlier, political communities ought to be understood as a communities of fate; the existence of a community is premised by a shared condition, thrown as it were into a particular location in time and place. Thus, it may be a common predicament – or the threat of a common danger – that binds people together as a community. When we take a look at the scope and nature of environmental problems like those of global warming and biological diversity against this background then we have to conclude that both contemporary fellow citizens as well as past and future citizens are partly vulnerable to the same dangers emerging from these problems. In this respect we may be seen as living in a common world, and thus share at least an overlapping frame of reference. However, as Honohan argues, at this level it may be more appropriate to speak of common concerns than of shared goods because the 'good' in question is necessarily defined in negative terms (Honohan, 2002, p. 152). Our responsibilities to future generations will therefore be of a different kind. After all, they do not answer our desire for recognition in a direct way as contemporaries do. Nevertheless, as we judge, we appear to respond to the potential judgement of future citizens as the judgement of those who are partly subjected to the same conditions.

On this particular point, an analogy can be drawn with Strawson's notion of the 'reactive attitude'. Strawson endorses this notion to indicate the intertwining of our moral practices with reactions towards 'others' that are not transparent to us and cannot be articulated in publicly specifiable terms. With this notion Strawson challenges the idea that the limits of our responsibility are determined by conscious personal choice, conviction and principle. He argues, on the contrary, that to hold an agent responsible for his actions is an expression of a reactive attitude. In our attribution of responsibility to ourselves and others we are lead by intersubjective rules, according to which some 'reactive attitudes' are more appropriate than others. In the examples Strawson espouses, those 'others' are living persons with which the agent interacts. If we consider the ideal of sustainability in this light, there are still others with whom we participate in common practices and hold responsible for the continuity of our practices. However, the beneficiaries of our behaviour – future generations – are located beyond the scope of our current practices. In itself this difference should not prevent us from applying the Strawsonian insight to the question of sustainability, because, as Burms reveals, our reactive attitudes play a vital role in our treatment of animals and the deceased as well. The theory of Strawson leads us to acknowledge, that in the end, the limits of our responsibility and the scope of our judgements remain opaque, because they depend on the reactive attitudes we hold towards each other as members of communal practices. Those attitudes are interwoven with our practices and our form of life in ways that are simply not open to us. This means that reactive attitudes and their related practices are not open to general justification. In contrast to beliefs and judgements attitudes are neither true nor false, and are not warranted by anything over and above their standard conditions of applicability. The fact that we cannot give publicly specifiable reasons for doing one thing as opposed to another does not mean that anything goes. We are still left with choices about whether to respond to those who spoil the life conditions of future generations in a reactive way – with blame or indignation – or to understand their behaviour in objective terms thereby discharging the agent from future responsibility (Strawson, 1974; Burms, 2000; cf. Magill, 2000).

Conclusion

To conclude, I would like to give an overview of my neo-republican view on future responsibility. The neo-republican ideal of sustainability requires us to take responsibility for the continuation into the future of the civil practices in which we are currently involved. As we have seen, an integral part of our participation in civil practices is paying particular care to its future. Our present commitment brings with it a personal claim that takes possession of a particular future of indeterminate length. However, our political judgements in public discourse are warranted by contemporaries only. Those who do not yet exist can neither answer nor mirror our desire for recognition. And in the absence of any answer, a lasting desire for their recognition as discourse members will simply not emerge. Therefore, anticipated communication with future generations will remain illusive and cannot be a source for political judgements. This conclusion is underlined by the widespread experience that advocates of future generations either veil their own interests behind this rhetoric, or introduce a highly metaphysical, almost religious authoritative argument that cannot be a proper subject in public discourse. This is not to say that future generations have no role at all in political judgement and discourse on environmental concerns; we cannot include future people in the 'anticipated communication' inherent in political judgement as anticipated communication partners, but their claims could be among our topics of discussion. When we recognise that claims on behalf of future generations will make demands on us, we do not anticipate communication with them. Moreover, the 'call' we experience emerges from our practical involvement in those public practices, projects, movements and traditions that we are committed to at present. Our personal investment in these practices and our dedication to its purposes imply trust in an open time horizon.

Imagination plays a vital role in political action. As we consider the implications of our current practices for future generations, we will imagine a particular future world, but rather than calculating what would be in the interest of its inhabitants – future people – we have to judge for ourselves: what kind of world would we want them to live in? Which practices are worth carrying into the future? Which world would be worth remembering? By taking responsibility for the world we want to leave behind, and by safeguarding the continuity of our worthwhile practices, we do not need to hide behind hypothetical rights of future people. Judged from a neo-republican perspective, that would be a means to run away from our responsibilities. Future generations are only of indirect importance when it comes to this approach of future responsibility. Even if we claim that we anticipate their possible standpoint, our judgement is in fact a function of our present commitments. Underlying this judgement is our striving for continuity of what we care for. The strong conviction that future generations are entitled to a dignified existence similar to ours is derivative of this primary judgement. In order to distinguish this existential perspective from the perspective of sustainable development, I choose not to speak of an ideal of sustainability, but instead, of an ideal of continuity.

In searching for a common ground or foundation that renders the ideal of continuity an obligatory force, I found that we cannot rely on transcendental conditions underlying our practices as a language community nor can we rely on the idea of a *telos* inherent to the realisation of mankind throughout history. Rather, our commitment to survival and striving for continuation of common practices should be perceived as our response to a call that originates from the perspective of an infinite and open future; a future that stretches out in front of us, and invites us to engage in future considerations. This perspective functions as a 'constitutive outside': an elusive other to which we answer in a way to which we will never be able to fully do justice. There is always something that escapes our present judgement and action. We are seized by this 'surplus', and we act in response to this, but will never be able to give a full account of our action in terms of reasons, precisely because our action assumes a particular responsiveness that is not transparent to the agent. However, before we take refuge in speculative metaphysics, I would like to stress that our particular response to this call is warranted by what *we* think and do within the intersubjective practices of the language community of which we are a part. Rather than being a matter of introspection or future speculation, we appeal to one another in our response to the infinite future that thrusts itself upon us. In this response we express how we experience ourselves as human beings and what we really care about. In short, we answer this call, not as an individual but as a discourse community. And again, the importance of a public space and discourse is highlighted. Within the public arena, we hold each other responsible for striving after the continuity upon which we implicitly agree. The continuity claim of the others we live among is not something we can easily shake off, for who wants to appear as a blunt egoist or moron in front of those others who also constitute who we are?

As such, the ideal of continuity (that requires us to take responsibility for the continuation of the civil practices we are involved in) can appear as an intersubjective *ground* for environmental action and judgement, without being elevated to the ultimate status of a metaphysical *foundation*. Given the fact we are intersubjectively situated in civil practices, we can make no judgements other than by including future considerations. This is an anthropological necessity, inherent to human action, rather than a logical one. We do not have the sufficient and necessary arguments for obliging others to consider the future implications of their judgements and actions. Conceived in this way, we can hold on to the expression that we judge or act in favour of future generations, while maintaining that we make our judgements in dialogue with contemporary others and derive motivation from our interaction with contemporaries. Even if we sense that we anticipate recognition of future generations, this desire is buttressed by present judgements of contemporary others. In my view, future generations are neither the source, nor object of environmental responsibility. The source of environmental responsibility resides within our current practices, while the focus or object of responsibility is the continuation of those practices, rather than its imagined participants.

2.4 EDUCATION FOR AN OPEN FUTURE

If environmental citizenship implies an existential and active responsibility as neo-republicans and civil society theorists argue, then the question arises as to how people become responsible citizens in the first place. How are we prepared for participation in political life? Human beings are generally endowed with the gifts of speech and action, as Aristotle claimed when he characterised human beings as *zoon politikon* – political animals – though obviously we are not born as fully-fledged citizens. Therefore, new-born generations will always be in need of some form of civil education or initiation into political life. There are, however, various views on the general purpose and outline of such a civil initiation. In general, liberal theorists tend to prefer a more or less isolated curriculum of competencies required by the status of citizenship. In this view, civil education is mainly designed to communicate and inculcate a common core of knowledge, values, skills and attitudes that is regarded preresquite for participation in the political system of a particular country. Communitarians, on the other hand, stress the acquisition of virtues within the moral community in which children grow up and learn to participate before they become citizens. In their view, civil responsibilities naturally emerge from the responsibilities of community membership. A neo-republican approach to environmental education as we have just outlined would stand in opposition to both approaches. Contra mainstream liberalism neo-republicans would reject the possibility of defining necessary conditions antecedently required by democratic citizenship in terms of competences. Therefore, they would probably object to the idea of an isolated education of civil competence as liberals propose. But this rejection would not lead them to adopt the communitarian approach, because participation in a moral community can never serve as a learning experience for participation in civil society. After all, the *political* community is profoundly marked by plurality instead of moral homogeneity, and requires a different attitude and way of judging.

Neo-republican thinkers and civil society theorists do argue that a particular form of moral bond exists between members of a political community. According to Hohohan the republic is 'a specific community of citizens related by substantial ties and a sense of loyalty more like fraternity or friendship than agreement on institutions or procedures' (Honohan, 2002, p. 6). This loyalty is not due to a common way of life or a common morality, but, as indicated earlier, the political community can be understood as the expression of a common fate. Oldfield for instance borrows the ideal of 'concord' from Aristotle in order to describe this particular kind of solidarity:

> *Concord is, thus, that friendship which exists between citizens as members of a political community. It is a relationship between people who know each other; they are not strangers, between whom goodwill is possible, but not friendship. It is a relationship between people who differ in their talents and capacities. It is a relationship based on respect for such differences, and on concern for others' interests; each, thus, acknowledges the other's autonomy. It is above all a relationship based on recognition that living is a shared venture, that can only be successfully engaged in if there is commitment. The commitment, however, is to the fellow-citizens, who – in choosing amongst themselves how to conduct their shared lives in the spirit of justice – create and sustain a community* (Oldfield, 1990, pp. 122–23).

Because of their trust in the strength of the political community some neo-republican thinkers and civil society theorists place their faith in the educative power of the process of political participation itself. Inspired by the ideas of Rousseau and J.S. Mill they assume that political participation itself will elicit civil virtue and teach people responsibility and toleration. In this view, the more one participates in public discourse the more one develops the attitudes and public spiritedness appropriate to a citizen. As Oldfield notes, they foster hope in the activity of participation 'as the means whereby individuals may become accustomed to perform the duties of citizenship. Political participation enlarges the minds of individuals, familiarises them with interests which lie beyond the immediacy of personal circumstance and environment, and encourages them to acknowledge that public concerns are the proper ones to which they should pay attention' (Oldfield; cited in Kymlicka and Norman, 1990, p. 361)[19]. Furthermore, those neo-republican thinkers expect that political participation of some will inspire others to join them: 'the example set by the initial participators will draw ever widening groups of individuals into the political arena' (Oldfield, 1990, p. 155; cf. Van Gunsteren, 1992).

Oldfield criticises this faith in the educative power of participation for being overly optimistic. He perceives their advocacy for collective participation and decision-making as a means of resolution to all kinds of social problems as a way to bypass the issue of responsible citizenship. In his view, the 'moral character which is appropriate for genuine citizenship does not generate itself; it has to be authoritatively inculcated. This means that the minds have to be manipulated. People, starting with children, have to be taught what citizenship means for them, in a political community, in terms of the duties it imposes upon them, and they have to be motivated to perform these duties' (Oldfield, 1990, p. 164). Oldfield goes on by discussing educative proposals for national service – civilian as well as military – which are expected to install a public spirit within the new generation entering our political community. Obviously, these particular remedies to fulfil the need for civil education will not be shared by those who approach this question from a progressive-educational point of view or those left-wing participatory democrats and civil society theorists who equate learning with participating.

When it comes to the educational issue, there are huge differences of opinion among neo-republicans and civil society theorists. But what appears to unite them is the fundamental conviction that civil education should take place within the public sphere itself. The (re-)creation of citizenship is itself regarded as a core responsibility of the political community. Citizens are expected to organise their political life and structure their institutions in a way that sustains and preserves the public spirit and engagement in collective affairs of each and every (future) citizen. Therefore, neo-republicans argue that this collective responsibility cannot be left to the moral community or to individual choice (Van Gunsteren, 1992, p. 119; Kymlicka and Norman, p. 361). In line with this neo-republican perspective, environmental education should be understood as a particular kind of initiation into our civil practices; an initiation in which newcomers are gradually involved as fellow-citizens, who help to shape these practices and are, *ipso facto*, gradually assuming co-responsibility for the continuity of our practices in the future.

But what exactly is it, that educators do when they involve children in our practices and simultaneously strive for continuity of those civil practices? Are environmental educators merely offering an appealing perspective on a possible future world, or do they treat their pupils as human material for the establishment of this future world to come? Derek R. Bell testifies to the latter by stating that the aim of environmental education ought to be to "create a new generation of citizens who are greener than their parents" (Bell, 2004, p. 43). In others words, does the claim of continuity do justice to the right of children to an open future horizon, or does it lead to a colonisation of our future in such a way that we deny new generations the opportunity to shape their own future world? This latter danger is precisely what the German philosopher of education Theodor Litt described as the 'Vorwegnahme der Zukunft'. Apart from Litt, philosophers of education generally warn against the utilisation of education as a means for creating some preconceived 'ideal society'. On the other hand, every educational act and judgement casts its shadows ahead. Every educational intention or response lays claim to a particular future of indeterminate length. By acting in a particular way now, the educator commits himself and the child to doing something in the future. Moreover, both our action or inaction has consequences for the world we leave behind for the next generation. Therefore, we cannot remain innocent and complete refrainment from continuity claims in education cannot be logically maintained. Thus, on the one had tension exists between the inevitability and constituting force of the promise of continuity and the risk of treating the present as a technology of the future on the other (Meijer, 1996).

In my view, this particular dilemma originates from the paradoxical nature of educational responsibility, as characterised by Arendt in her notorious paper *The Crisis of Education* (1958). Educators bear a double responsibility, so she argues, one of which aims to preserve the child's openness and newness against the powers of the established world, whereas the other aims to protect the durable world against the potential recklessness of the new generation:

> *In education they (adults – DWP) assume responsibility for both, for the life and development of the child and for the continuance of the world. These two responsibilities do not by any means coincide, they may indeed come into conflict with each other. The responsibility for the development of the child is in a certain sense a responsibility against the world: the child requires special protection and care so that nothing destructive will happen to him from the world. But the world, too, needs protection to keep it from being overrun and destroyed by the onslaught of the new that bursts upon it with each new generation* (Arendt, 1958b, p. 504).

This tension is prevalent in the practice of environmental education because here, in particular, both child and world are vulnerable to each other's powers. On the one hand, every new generation carries with it the promise of a new future and the ability to think about worldly problems anew and to go in a direction previously unthought of. This is an expression of what Arendt calls 'the condition of natality'[20]. In her eyes, it is the task of educators to preserve this newness over and against the established powers that may force them to act and think about the world and the environmental crisis in a particular preconceived way, for instance, as a problem of resource management. On the other hand, every new generation constitutes a potential threat to the durability of our world, and it is the task of educators to take responsibility for this world in front of them. They are to protect the environmental goods of our world against destruction and decay, for example, against the consumerist threats inherent in contemporary youth culture. In this sense, Arendt argues, education is and should be a conservative enterprise:

> *Basically we are always educating for a world that is or is becoming out of joint, for this is the basic human situation, in which the world is created by mortal hands to serve mortals for a limited time as home. Because the world is made by mortals it wears out; and because it continuously changes its inhabitants it runs the risk of becoming as mortal as they are. To preserve the world against the mortality of its creators and inhabitants it must be constantly set right anew. The problem is simply to educate in such a way that a setting-right remains actually possible, even though it can, of course, never be assured. Our hope always hangs on the new which every generations brings; but precisely because*

we can base our hope only on this, we destroy everything if we so try to control the new that we, the old, can dictate how it will look. Exactly for the sake of what is new and revolutionary in every child, education must be conservative; it must preserve this newness and introduce it as a new thing into an old world, which, however revolutionary its reactions may be, is always from the standpoint of the next generation, superannuated and close to destruction (Arendt, 1958b, p. 510).

In Arendt's view adults are able to fulfil their twofold task of protecting the young against the world and the world against the young only by introducing the young into our world as it presently is, in all its potential and with all its flaws. Educators stand in relation to the young as representatives of the present world for which they must assume responsibility even though they themselves did not make it, and even though they may, secretly or openly, wish it were other than it is. As such, educators must present an authentic picture of the world that is open to critique and change. Arendt therefore argues against those progressive educators who introduce the young into a desired world instead of the world we actually live in. It would be a big mistake to transform our picture of the world on behalf of the young as though our desired future were already a reality: 'These educators unwittingly send the message to students that the world is no longer in need of transformation; it has already been transformed' (Levinson, 1997, p. 443). For this reason, educators should not inculcate us with the ideal of a bright green future *Ecotopia*. Instead they are called upon to show what commits us to the present world. Rather than being utopian prophets, environmental educators ought to be guardians of our common world. Moreover, they ought to familiarise children with the problems that threaten our common world as part of their introduction into that world, so Arendt argues. Only when educators refuse to take responsibility for these worldly problems, and take refuge in utopias, do risks of indoctrination and manipulation lie in wait (Arendt, 1958a, p. 307; Achterhuis, 1996; Levinson, 1997; Gordon, 1999; Vansieleghem, 2004).

Arendt refines her statement that education should be conservative by stressing that education should take place in 'the gap between past and future'. The present offers us a possible space in which to recollect and understand the past in a way that does not determine the future, but does throw an illuminating light on it, and fosters responsibility for the future. As representatives of the present world, educators mediate between the old and the young by balancing between that which is 'no longer' and that which is 'not yet', or, as Levinson notes: 'To teach in this gap is to take on the twofold task of introducing students into a world that precedes them, while preserving the possibility that students might undertake something new in relation to this world'. And: 'to teach in this gap is to commit ourselves to teaching about the past – for understanding and guidance, and for the preservation of memory that underlies both – and to motivate students to try to set things right. At the same time we have to resist the temptation of attempting to determine and control our students' futures' (Levinson, 1997, p. 450). According to Levinson this gap is not mainly to conserve and preserve the past, but occupies a provocative space as well, one which opens the possibility of interrupting social processes that appear fixed and inevitable (Arendt, 1958a; Gordon, 1999; Levinson, 1997; Vansieleghem, 2004).

At a recent conference on environmental education the Dutch state official Herman Wijffels posed the rhetorical question: 'Do we educate pupils to cope with the problems of today or do we educate them for answering the challenge of tomorrow?'[21]. If I were to answer this question – despite its rhetorical nature – I would agree with Arendt and prefer to educate children for dealing with 'the problems of today', because those who pretend to know 'the challenge of tomorrow' are either reformulating present problems in terms of future challenges by invoking the speculative authority of future generations or they are taking possession of an indeterminate future that should remain open for posterity.

Perhaps it is due to the metaphysical nature of Arendt's ideas on educational responsibility, that her advocacy of conservatism can so easily be misunderstood. What does Arendt precisely mean when she states that educators must assume responsibility for the present world in which they introduce the young? Does her statement imply that environmental educators are to justify the present *status quo*? Does she mean that they have to render the existing institutions, practices and procedures legitimacy by providing the arguments that sustain their *raison d'être*? Obviously not. Educational responsibility is not primarily concerned with justification or legitimisation, but with responsiveness. In their words and actions educators have to *respond* to a common world, rather than turn their backs on this world and creating an educational 'micro-world' in which they hold the young captive. By responding, educators introduce the young into a common world, as if they want to say: 'This is our world'. Only by introducing new generations to our common world and its practices, will they be able to see its strengths and weaknesses, its beauty and ugliness, its goods and bads. Only then will they be sufficiently prepared to participate in public discourse, to speak and act in order to change or protect these practices[22]. This conclusion is in line with the analysis of the preceding in which our future responsibilities are derived from those practices we care for here and now.

Once this is clear, the question remains as to how educators raise awareness of and instil commitment to what we experience as good and meaningful in our world. In order to understand the nature of our allegiance to common practices and rules, we should dwell for a moment on the ideas of Mouffe, who argues that 'allegiance to democracy and belief in the value of its institutions do not depend on giving them an intellectual foundation' (Mouffe, 2000, p. 97). In the end, fostering allegiance is not a matter of giving reasons but of persuasion against a background of shared beliefs. On this particular point, Mouffe elaborates on the later work of Wittgenstein in which he stresses that to agree on opinions, there must first be agreement on the language used, and agreement in the use of language depends on agreements in forms of life: the inescapable horizons of certainties on which we rely in our daily actions by sharing a common life. These 'certainties' are not grounded in reason but rest on trust or belief. For Mouffe, allegiance to democratic practices is similar to what Wittgenstein calls 'a passionate commitment to a system of reference. Hence, although it is *belief*, it is really a way of living, or of assessing one's life' (cited in Mouffe, 2000, p. 97).

This commitment is not warranted by rational consent, but established by means of identification with the beliefs we experience as 'true' or 'good' in our daily practices, and through recognition of the demands these beliefs make on us. Following the later Wittgenstein as well, Rorty underlines that to call somebody irrational in this context 'is not to say that she is not making proper use of her mental faculties. It is only to say that she does not seem to share enough beliefs and desires with one to make conversation with her on the disputed point fruitful' (cited in Mouffe, 2000, p. 65).

In line with these Wittgensteinian insights, Ramaekers and Smeyers (2004) argue that education should be seen as a particular kind of initiation into a 'form of life' understood as a horizon of beliefs:

> *What Wittgenstein envisages here is, from the perspective of the parent, initiating the child into a picture of the world, into a shared practice of doing things this way and not the other. For the child, it is about acquiring a pattern of beliefs – belief not in the sense of "lacking the proper grounds and therefore merely belief instead of knowledge", but belief understood as a foundational expression (Cf. Stroll, 1994). The child is acquiring the very grounds or "substratum" (Wittgenstein, 1969, # 162) of what is understood to be a human existence. This is shown primarily in the way she acts. She does not learn these beliefs propositionally – they may never have been expressed, thus Wittgenstein – but learns to act accordingly (Cf. Wittgenstein, 1969, # 144), "learns to react in such-and-such a way" (Wittgenstein, 1969, # 538). To be precise, this is not a matter of acquiring single beliefs or propositions, but of coming to believe "a whole system of propositions" (Wittgenstein, 1969, # 141), of which Wittgenstein says in the very same section, that "Light dawns gradually over the whole". Importantly, the giving of reasons to bring in some kind of epistemological justification is not what is involved here. Rather, trust, in its most basic form, is what takes centrestage here* (Ramaekers and Smeyers, 2004).

Wittgenstein's point is precisely that the exchange of reasons comes after the 'belief' or 'conversion'. Accordingly, the authors suggest that education might be thought to consist of the persuasion of children to 'see things differently'. Not by arguing for or against a particular perspective, but by gaining trust in the practices in which we introduce children, we invite them to see things from this or that particular angle. Sometimes, it is by seeing a particular aspect of reality in a different way that the entire picture of reality changes (in this context Wittgenstein discusses what happens to us when we see the well-known picture of a duck that suddenly transforms into a picture of a rabbit as soon as we identify that which we took to be the duck's beak as the ears of a rabbit. Here, Wittgenstein speaks of 'the dawning of an aspect'). This insight in what might be labelled a gestalt switch can be extrapolated to an ethical and esthetical context as well. Thus, the shift of an anthropocentric view of nature to a more poetic world-view that leaves room for recognising beauty, integrity or value in nature, might indeed be 'triggered' by seeing this particular tree from a different angle; as a powerful symbol of strength and vitality, for instance, rather than an irritating obstacle on the sidewalk. To give a more practical example: if city children are taken on a field trip to an organic farm in the countryside to study and work in the field, then this involvement i.e. experience summons them to see their daily food in a different way; not as consumer commodities you buy in a supermarket or eat in a fast food restaurant, but as the fruit of our land that sustains our very being. Raising an awareness like this – an awareness of the symbiosis we enjoy with our land – cannot (merely) be passed on to others in terms of rational arguments and propositions. Moreover, such an awareness is part of a belief system or a picture of the world, that cannot be argued for or against. An awareness like this can only be evoked by practical experience with which educators familiarise children by introducing them into such a caring practice (Wittgenstein, 2001, p. 295; Ramaekers and Smeyers, 2004).

To sum up, educators foster allegiance to their common practices and the underlying rules governing our behaviour within those practices by persuading children to participate; by inviting them to act in agreement with those rules and to live with the underlying truths and certainties as we do. An initiation of this kind will enable children as newcomers in our civil practices to apply rules in a different way than we commonly perform, and as such, change the rules and meanings of our practices. Furthermore, an initiation like this will render new participants receptive to the evolution of the background beliefs and shifting truths in time. Consider, for example, our growing sensitivity to animal welfare, our intuitive recognition of particular future claims or the sense of beauty we experience in our involvement with the natural environment. Against this background four features emerge as constitutive for the form of environmental education we have in mind: the focus on practice, plurality, conflict and double responsibility.

Practice

Since it is impossible to detach environmental knowledge, values, skills and attitude from the practices in which they are expressed and acquire meaning, education must consist of an introduction into meaningful practices as a whole, rather than an isolated education of action competences. Therefore, environmental education is not primarily concerned with the transmission of environmental knowledge, ecological values, skills and attitudes, so that the child can function in particular practices. Moreover, the educator is focussed on the selection and design of an ensemble of practices for an educational purpose: environmentally caring practices that foster particular kinds of identification and make possible particular kinds of individuality. The environmental educator looks for caring practices in which a particular involvement with the natural environment and care for the future is preserved and a particular model of ecological responsibility is practised. In such caring practices, children are familiarised with a style of interacting with nature and anticipating future concerns that is ultimately different from those interactions we usually employ as consumers. Thus, environmental education comprises an initiation into counter-practices that offer an alternative picture of the world, an alternative ethos and an alternative model for experiencing our contact with the natural milieu. The common ground for an environmental morality should not be located in a common rationality, but in a common practice plus its subsequent ethos and belief.

Plurality

Since it is impossible to define what is 'reasonable' without referring to certain pivotal judgements within paradigmatic practices – my devotion to the Sunday afternoon walk in a nearby forest, my dedication to American cars or my attachment to traditional farming practices – it would be unjust to resort to one conception of justice in our definition of the conditions of citizenship. Rather, there are many ways in which we can practice environmental citizenship within a civil society and we have no practice-transcendent reason to judge one above the other. Environmental educators and citizens are not concerned with applying general rules or values in particular cases, rather, it is the other way around. Particular cases will lead them to identify an 'ensemble of practices' each of which subsumes some aspects of our environmental responsibility. There is no 'correct understanding' of environmental responsibility that every rational being should be willing to accept but a plurality of ways to approach the environmental issues we are confronted with. In this light, Mouffe argues for a plurality of practices employed for the formation of citizenship: 'Democratic individuals can only be made possible by multiplying the institutions, the discourses, the forms of life that foster identifications with democratic values' (Mouffe, 2000, p. 96). In a similar way, environmental education should be understood as an introduction into an ensemble of environmentally caring practices, thereby persuading children to broaden the range of their commitments to others, to build a more inclusive community that comes to terms with the imperatives of future continuity and responsibility for nature. These practices are collectively labelled 'educational' and 'environmental' because they are related by means of family resemblances rather than subsumed under a uniform conception of 'sustainability' or environmental education.

Conflict

Since there is no master-practice that renders all practices their relative place within a larger framework of environmental responsibility, ambiguity and conflict will always be at the heart of environmental education. Every practice and potential articulation of our care for nature necessarily excludes other possible ways of acting and caring. There is no unifying framework beyond the sheer plurality of practices. This plurality does not only engender *interpersonal* conflict between citizens, communities and associations within civil society, as we have seen above, it also implies *intrapersonal* conflicts of loyalty and identification as well. The divergent loyalties and identifications children develop by participating in disparate practices sometimes harmonise within a coherent idea they have of environmental responsibility, but they may conflict as well. A familiar example is the conflict between our role as consumers and our role as world citizens when it comes down to the purchase of fair trade products. Whereas my vigilance when it comes to shopping and constraints not to exceed the household budget urges me to keep away from the expensive products of fair trade, my responsibility as a world citizen requires me to consider the social and ecological conditions under which the products are produced and will probably motivate me to buy the fair trade product instead of the cheaper alternative. But how should we weigh both claims against one another, and what kind of reflection or determination should settle a personal conflict like this? The exact nature of these conflicts and possible ways of dealing with them will be explored in the following chapter. According to Honohan, it is the split between private and public concerns that highlights 'a tension *within* each person between the immediately perceived particular advantage of each and the general interest of the citizen as an interdependent member of the polity, and requires each to be active in pursuit of the common good, to have *public spirit*, and to participate in public service' (Honohan, 2002, p. 158). Mouffe stresses that 'these conflicts are inevitable since our identities are shaped by a plurality of changing audiences' (Mouffe, 2000, p. 59). As such, dealing with conflicts of loyalty on the borderline between disparate practices is not only part of the process of becoming a citizen, but this is a necessary part of the process of increasing self-understanding as well. For this reason, I want to argue that, apart from the caring practices to which children are introduced, environmental education also requires a separate space within which pupils and student are enabled to deliberate on these conflicts and exchange experiences. A discursive practice of this kind should aim at making collective reflection on environmental responsibility possible rather than subsuming all practices under a common practice or framework of environmental education.

Double responsibility

Unlike the educational ideas of some neo-republicans, the analysis given above leads us to assume that the practice of citizenship does not reproduce itself without some educational guidance. Participation alone does not breed citizenship, since new-born generations need to be introduced into the civil practices in which they are thrown. Without education, the young would probably be lost in a civil society that is far too complex and diversified for them to find their way around by themselves. But even more important, the public sphere is profoundly marked by display and exercise of power against which newcomers are defenseless. Without adult guidance the existing powers would undermine the promise of young generations to bring along something new. They would lose their ability to speak and act anew in an old world. This need for a guided introduction calls on the double responsibility of environmental educators; by virtue of their position as mediators between an old world and new generations, educators need to be responsive to the implicit claims of young generations to an open future, as well as to the implicit claim of our old world in need of protection. In fact, educators maintain a relation based on trust with the young as well as the old; they require the young to trust the old and require the old to trust the young by protecting the newness of the young against the established powers of the old and protecting the established world against the destructive newness of the young.

In Arendt's view adults are only able to fulfil their twofold task by introducing the young into our world as it is at present, in all its potential and with all its flaws. Only when educators act as representatives of our present world, for which they assume responsibility, can the fundamental conflict between both claims be dissolved. This is because in order to preserve our common world it has to be continuously reformed, and reform is only possible when newcomers take their bearings from the present world in which they are properly introduced. According to the *Geistenwissenschaftliche* philosopher of education Klaus Mollenhauer, we educate the young so that the good in our life may continue to exist (Mollenhauer, 1986). In a similar spirit Hargrove claims that humans have 'a duty to promote and preserve the existence of good in the world' (Hargrove, 1992). So, rather than being utopian prophets environmental educators should be guardians of our common world. Moreover, Arendt also argues they are to familiarise children with the problems that threaten our common world as part of their introduction into that world. Only when educators refuse to take responsibility for these worldly problems, and take refuge in utopias, do risks of indoctrination and manipulation lie in wait (Arendt, 1958a, p. 307; Achterhuis, 1996; Levinson, 1997; Gordon, 1999).

Of course, the introduction of the young into our common world and its problems should proceed gradually. Arendt sharply delineates the line between the political practice of adults, acting among equals, and the educational practice in which adults cannot assume equal responsibility for the world among children (according to Arendt the Greek word *schole* literally means 'abstain from political action'). This insight in the discrepancy between adults and children leads me to argue that adults should not confront children with environmental responsibilities they cannot live up to themselves. Rather than as an educational issue, these responsibilities ought to be a subject of political action. As such, contemporary citizens cannot shirk their responsibility to future citizens (cf. Achterhuis, 1996). Furthermore, adults should not confront children with responsibilities for problems they cannot possibly oversee. Educators must introduce children into environmental problems only if and insofar as these practices offer them concrete possiblites for taking action. In the absence of possibilities for taking action, environmental educators risk bringing about destructive feelings of fear, impotence and defeatism. So, obviously, there is no use in teaching six year old children about acid rain, for they will not be able to grasp the problem and judge its implications in proper proportions. Experience has shown that a premature introduction into problems of acidification might indeed make children afraid of rain (When it once started raining a six year old boy remarked to his mother: 'The trees are all going to die, right now'). It might perhaps be more valuable to involve six year old children in our collective efforts to keep the playgrounds around school free of litter. Introduction into environmental practices is meaningful, only when children are able to apply the rules of these practices and when they are in a position to co-shape those rules and apply them in pursuit of a different future for those practices. As such, a gradual introduction into environmental problems and practices – following the extension of the children's' life–world – may inspire them to take care of their environment without losing trust in the continuation of our worldly practices.

Conclusion

The first section of this chapter ended with the conclusion that liberal theory, underlying present proposals for environmental education, presents us with a false dichotomy. Dependent on its success, the liberal attempt to justify sustainable development on second-order grounds either results in a wishy-washy domestication of environmentalism – treating ecological values and standards as mere lifestyle options – or it ends in practices of indoctrination – imposing sustainable development as a firm ideology upon (future) citizens. A viable way between the two extremes of complete permissiveness and indoctrination cannot be offered by communitarianism since its understanding of intergenerational responsibility is too limited in scope and the underlying assumptions concerning community life and personal identity are unrealistic and flawed. However, such a viable way may be found by adopting a neo-republican perspective, as outlined in this section. By focussing on practice, plurality, conflict and double responsibility as being the constituent parts of environmental education, children will then be inspired to take responsibility for the continuation of the practices they are involved in, without being pinned down to a particular ideology or future condition of this practice.

To be more precise about the contribution of our neo-republican perspective on environmental education: the focus on *practice* prevents us from rationalising our contact with nature in a way that turns this relationship into a mere option for choice or, on the other hand, into an ideological relationship. Moreover, the focus on practice requires us to participate in those practices within which our care for the natural environment will be preserved in a way that dovetails with our self-understanding. The focus on *plurality* prevents us from adopting one single practice as a model for environmental responsibility, but forces us to be receptive to those perspectives that are excluded by the current practice in which we participate. The focus on *conflict* prevents us from striving for an agreement that settles for only one particular image of the future as being worthy of our attention and simultaneously prevents us striving for a free-floating agreement that leaves everything to voluntary choice. Moreover, the focus on conflict opens up the possibility of substantive debate in which we call on one another's responsibility. And, finally, the focus on *double responsibility* prevents us from sacrificing the freedom of the young to the powers of the established world, or sacrificing the powers of the established world to the freedom of the young. Moreover, their double responsibility requires environmental educators to be receptive to both the young and the old whose claims appear to join in the present necessity to act and respond to the dangers that threaten our perspective of an open future.

NOTES

[1] Arendt, 1958b, p. 510.

[2] A similar analysis is posed by Derek R. Bell, who states that political liberalism requires the curriculum of all schools to meet the so-called JBUC-standard: 'It is important to notice that the requirements that political liberalism imposes on educational content apply to all schools whether they are state schools, voluntary-aided schools or private schools. 'Society's concern is with [children's] education

lies in their role as future citizens at all types of schools' (and those educated at home) are future citizens (Rawls, 1993, p. 200). We might call these requirements '*j*ustice-*b*ased' (to acknowledge their importance in the maintenance of just institutions over time through the social reproduction of the political virtues), *u*niversal (to acknowledge that they apply to all places of child education) and 'compulsory'. For short, the JBUC curriculum (Bell, 2004, p. 39).

[3] I also made use of the revised edition of *A Theory of Justice* (1999) in which Rawls revised some of his original ideas on the issue of intergenerational justice.

[4] Barry's adjustment of Rawls' theory of justice has explicitely been adopted by Achterberg (1994b), Hilhorst (1987), Van der Wal (1979) and implicitly by many others.

[5] In section 2.3 it will become clear that it is inadequate to seperate the content from the conditions of freedom in this way. However, at this point I judge the credibility of the argument within the Rawlsian framework.

[6] This formulation of the non-identity problem is borrowed from de-Shalit (1995).

[7] The terms 'goods' and 'bads' are not used by Rawls; they are my own.

[8] As we will discuss latersuch, the liberal contract aggregates private choices, but ignores the community-shaping function of deliberative judgments (cf. Whiteside, 1998).

[9] In Rawls' view the issue of moral consideration for animals is beyond the limits of a theory of justice: 'While I have not maintained that the capacity for a sense of justice is necessary in order to be owed the duties of justice, it does seem that we are not required to give strict justice anyway to creatures lacking this capacity. But it does not follow that there are no requirements at all in regard to them, nor in our relations with the natural order. Certainly it is wrong to be cruel to animals and the destruction of a whole species can be a great evil. The capacity for feelings of pleasure and pain and for the forms of life of which animals are capable clearly imposes duties of compassion and humanity in their case. I shall not try to explain these considered beliefs. They are outside the scope of a theory of justice, and it does not seem possible to extend the contract doctrine so as to include them in a natural way. A correct conception of our relations to animals and to nature would seem to depend upon a theory of the natural order and our place in it. One of the tasks of metaphysics is to work out a view of the world which is suited for this purpose; it should identify and systematize the truths decisive for these questions. How far justice as fairness will have to be revised to fit into this larger theory it is impossible to say. But it seems reasonable to hope that if it is sound as an account of justice among persons, it cannot be too wrong when these broader relationships are taken into consideration' (Rawls, 1999, p. 448–449).

[10] In general terms the so-called 'prisoner's dilemma' – formulated within early game-theory – can be formulated as such: in situations in which the realization of a common good requires sacrifices of *all* individuals of a particular group or society, the presumed calculative c.q. rational-economic bias of individual behaviour will discourage individuals to cooperate, since the benefits of each individual cooperation are uncertain and spread out over the collective, while the benefits of non-cooperation are certain and individually enjoyable. Ergo, the structure of this type of situation rewards the non-cooperative behaviour of the 'free-rider', who benefits from the sacrifices of fellows without sacrificing herself. Generally speaking, two types of solutions are proposed to overcome this dilemma. First, there is the informal 'tit-for-tat-startegy' designed to realize self-regulation within small-scale groups; all individuals freely promise to act cooperatively. Second, in large-scale, impersonal situations there is the solution of 'mutual coercion mutually agreed upon', which comes down to the formalization of force: the enforcement of cooperative behaviour by means of sanction mutually agreed upon. This solution is often assumed to require a central institution – for example the state as 'penal machinery' (similar to the Leviathan of Hobbes) – that oversees compliance. Both strategies consist of providing assurance of fellow-cooperation, so that the barriers for cooperation are removed (the fear of free-riding fellows is then minimalized (cf. Van Asperen, 1993).

[11] As such, my strong conception of the civil society is distinguished from the liberal conception that regards the civil society as synonym with the free market or communitarian conceptions that conflate the civil society with the moral community (cf. Barber, 1997 and 1998).

[12] NIMBY groups, as they are called by their critics, are those for whom change is tolerable so long as it is *Not-In-My-Back-Yard* – that is, so long as it does not affect members of the groups themselves.

[13] This is my term, borrowed from Van Middelaar, (1999). *Politicide*. Amsterdam: Van Gennip.

[14] Borrowed from the Treanor website : 'Why sustainability is wrong' (http://web.inter.nl.net/users/Paul.Treanor/#lib). On his site Treanor notes a conversation with an advocate of sustainability: 'What is your position on those who oppose sustainability?"/ "No- one is against sustainability"/ "I am"/ "No, I think we just disagree on the definition"/ "I am against all definitions"/ "No-one can be against sustainability". Then Treanor somewhat cynically remarks: "This is probably inherent: if objections of conscience are recognised, no sustainability policies are possible. It is like belief in a universal God: if you believe God is only for believers, then you do not believe in a universal God. By nature, sustainability must claim a monopoly of belief: as a "belief" it cannot admit an opposite belief is equally valid. It is a consistent and Universalist world-view, Weltanschauung. Its adherents act in accordance with one general principle: that it should be accepted by all persons'.

[15] This is not to say that procedures are seen as redundant and should be pushed aside. On the contrary, formal procedures do indeed contribute to justified judgements in complex situations. However, Mouffe's point is that these formal procedures as such do not warrant justice because they have to be applied in social practices that are profoundly marked by conflict and plurality, which means that there is not necessarily agreement among the parties on the status and nature of the conflict at hand and the kind of procedures that the conflict asks for. So, in the end, justice depends on a particular sensitivity, trust and prudence that allow the parties or arbiters to apply the procedures in a just spirit. Conclusively, the disagreement between Mouffe and Rawls is not on the necessity of procedures but about the status of procedures: for Mouffe they are part of political practice as such, whereas Rawls renders them a meta-status: the procedures are seen as preceding political practice and securing justice within this practice.

[16] Arendt substitutes Kant's imperative of universalizability for an imperative of generalizability (Nieuwkerk and Van der Hoek, 1996).

[17] Arendt points out that Kant's idea of progress conflicts with his notion of human dignity, that requires us to respect human beings in their particularity, without treating them as representatives of (the idea of) humanity (Arendt, 1982, p. 118).

[18] In the introduction to the so-called *ESD-toolkit* a similar insight is articulated through an argument against those who are waiting for a perfect definition before they consent to collective action: 'It is curious to note that while we have difficulty envisioning a sustainable world, we have no difficulty identifying what is unsustainable in our societies. We can rapidly create a laundry list of problems – inefficient use of energy, lack of water conservation, increased pollution, abuses of human rights, overuse of personal transportation, consumerism, etc. But we should not chide ourselves because we lack a clear definition of sustainability' (http://www.esdtoolkit.org/discussion/default.htm).

[19] It should be noted, however, that Arendt herself was not in favour of a civic education within the public sphere itself. However, her placement of educational practice within the social sphere gave rise to an extensive debate about the precise delineation between the social sphere and the private as well as the public sphere (that we will not discuss right here, since a more close study of Arendts thoughts on this point would draw the attention from the core issue in this section). Nevertheless, we can conclude that Arendt did regard school education as a responsibility of the community rather than a private matter of parents (Arendt, 1958a, 1961).

[20] Arendt writes about natality as the condition of human action and, *ipso facto*, of human freedom: 'action has the closest connection with the human condition of natality; the new beginning inherent in birth can make itself felt in the world only because the newcomer possesses the capacity of beginning something anew, that is, of acting' (Arendt, 1958, p. 176). Thus understood, Arendt's idea of natality is very similar to Lyotards notion of childhood – *l'enfance* in terms of a radical indeterminacy that cannot be represented (Smeyers and Masschelein, 2000, pp. 149–152).

[21] This is my translation of Wijffels' question, that was originally posed in Dutch as follows: 'Leiden we leerlingen op voor de problemen van vandaag of voor de uitdaging van morgen?'.

[22] Arendts' expression of educators' double responsibility and her advocacy of conservatism can be understood as a response to the famous pedagogical paradox, as expressed by Kant: 'How do I cultivate freedom through coercion?' (*Wie kultiviere ich die Freiheit bei dem Zwang*). In other words: 'How can a free, human being, who, at birth, is not what he should be, be prompted by external means (i.e. by an educator) to become what he should be?' (Vanderstraeten and Biesta, 2001, p. 10).

CHAPTER THREE

BECAUSE WE ARE HUMAN

BWA-PL

After a bike ride through a dripping wood
we reached
the Border Lake.
It was as if a sleeper opened her eyes
and knew us
You sat in front.
I laid my hand
On the warm coconut of your skull
The light looked far into your eyes
I said: now this is water.
Wa-ter
Wa-ter
Wa-ter, I said again
And you said: bwa-pl
You said it again
It was certain, son of mine, that it was the same
we did not understand

Willem Jan Otten[1]

If future responsibility emerges from our commitment to what we care for here and now, rather than our hypothetical relationship with future citizens, how do we stand towards the things we value? Do we 'make' our natural environment valuable by ascribing value, or do we 'find' value in nature? Perhaps both acts of ascribing and finding value are operative at the same time, but if this were so, how do the acts of ascribing and finding value relate to one another? These questions will be at the heart of this chapter. Within the field of environmental ethics there is a long history of debate between those who defend that natural value resides within the subject – the evaluator – and those who argue that the value is intrinsic to the object – nature itself. Most environmental philosophers, however, place their position somewhere in between the extreme ends of value-subjectivism and value-objectivism. In the first section of this chapter, section 3.1, we will explore where the advocates of Education for Sustainable Development (ESD) stand with respect to these questions of natural value. There is much debate within these circles on 'the intrinsic value of nature', but it is seldom clear what the proponents imply by claiming intrinsic value. Therefore, an index will be given of possible positions on the question of what constitutes the intrinsic quality of natural value. In general, all positions seem to imply that we value nature not merely for instrumental reasons, but the value of nature, in a sense, transcends our private interests and preferences. As such, a cow is more than a milk machine and a tree is more than a bearer of fruit or producer of oxygen.

Still, one might wonder why we are so deeply concerned with the question of intrinsic value. After all, ill-defined claims of intrinsic value are often a source of mystification, disguising inherent contradictions and controversies as to its meaning. Then why do we bother to enter into this debate? To begin with, we are not so much concerned with the question of intrinsic value itself, but we are interested in the core question of responsibility. Intrinsic value claims can be understood as expressions of responsibility. By expressing the things in life we really care for, we commit ourselves to preserve their meaning and pursue their good. For example, if someone declares her love for cats in public, she takes responsibility to care for them, otherwise the public will hold her responsible for the promise inherent in her testimony. Thus, examining the nature and meaning of intrinsic value claims allows us to examine the way we stand towards the things we value and the way we in which we take responsibility for their preservation and prosperity.

In the second place, our relationship to nature might be of an exclusive kind, in the sense that it requires us to value and act on its behalf in a particular way, different from our responses to human beings, communities, ideas or cultural products. As such, the status of environmental ethics as an independent field of study is often thought to be based on the claim of the intrinsic value of nature. If nature were conceived as a mere stock of human resources, as some kind of standing reserve for human consumption and exploitation, there would be no need to think about our relationship to nature in an ethical way. Or to quote Rolston: 'Environmental ethics in the *primary* (...) sense is reached only when humans ask questions not merely of prudential use but of appropriate respect and duty (towards the natural environment)' (Rolston III, 1988, p.1; cf. Regan, 1992, p. 161). In a similar way the status of environmental education as an independent educational practice rests on our recognition of the natural environment as intrinsically valuable. Without this recognition we would have no reason whatsoever to pass on our concern for nature to the next generation.

In light of this I will argue in section 3.1 that there is a sensible way of speaking about 'the intrinsic value of nature' in terms of aesthetic judgement. In this view, the value is neither intrinsic to the subject, nor to the object of valuation, but resides in the common language that enables us to speak about nature. As such, the value of nature is embedded in the intersubjective level of our language community. However, as I will argue, the intrinsic value of nature transcends our articulation of value. Its ultimate value and meaning eludes us. In section 3.2 I will examine in more detail the status of transcendent nature as valued object and the human agent as valuing subject within an intersubjective-aesthetic account of natural value. As we will see, the issue of intrinsic value is important to our understanding of what it is that inspires us to care for nature. How to understand the caring responsibility we assume in our practical involvement with the natural environment will be explored in section 3.3.

Nel Noddings' phenomenology of care as well as Maurice Merleau-Ponty's phenomenology of perception will be given precedence in my investigation of our corporeal experience of caring for nature that is immediately at hand. These phenomenological characterisations allow me to derive an ethical ideal of caring responsibility for the natural environment in section 3.4.

3.1 IN DEFENSE OF AN AESTHETIC ACCOUNT OF INTRINSIC NATURAL VALUE

For a long period of time, environmental discourse was marked by controversies between anthropocentric and ecocentric views on the nature and scope of the environmental crisis. As the word itself says, anthropocentrism places humans at the centre of reality, and views humanity as standing apart from and above nature. Or to be more precise, anthropocentrism 'involves the foundational assumption that human beings, and the way in which they value nature, are the *modus operandi* of any attempt to think about the green environment' (Smith, 1998, p. 4). Rather than a particular school of thought, anthropocentrism refers to a tendency inherent in western culture and philosophy, running from ancient Protagoras – who argued that man was the measure of all things – to the modern positivists who postulate nature as an object of human experiment and manipulation. In line with this tendency, anthropocentrists weigh environmental issues merely in terms of what is in the long-term interest of human beings, or in terms of what satisfies human needs and serves human purposes. Gifford Pinchot, for instance, a major figure in the American Conservation Movement of the nineteen forties and fifties, insisted that 'there are just two things on this material earth – people and natural resources'. As a provider of resources, necessary for human survival and prosperity, he regarded nature only of instrumental value. Environmental concerns are thus considered to emerge only in situations in which natural resources are scarce, or otherwise, in situations in which our resources are being threatened by forces of nature or by the unintended effects of human behaviour (Matthews, 2001, p. 227).

Anthropocentric analyses of environmental problems are pursued within the prevalent framework of the economic, political and socio-cultural structures in society. Generally, environmental problems are considered to be problems of control, manageable by employing the appropriate, economical and technological means with which these structures provide us. With respect to the issue of car use and exhaust fumes this means, for example, that the efforts of anthropocentrists concentrate on the installation of catalytic converters, the introduction of hydrogen cars, and perhaps, the inclusion of environmental costs in the fuel prices. However, the threats to nature that do not directly affect our life conditions are not on their agenda, neither are the structural questions concerning our modern ways of transportation, the economic interdependencies that regulate our transport behaviour nor, for instance the questions that have to do with the political power of the global oil industry.

Ecocentrism emerged out of dissatisfaction with this 'shallow' analysis of environmental problems. As a response, ecocentrist critics challenged the exclusive anthropocentric concern with issues of pollution control and resource conservation for the protection of people in the developed countries. By moving beyond the pursuit of human interest and assuming equal respect for all forms of life, ecocentrists aimed to extend the environmental agenda to issues of natural integrity, biological diversity and the preservation of biotic communities. Contrary to the adherence of anthropocentrism to the modern ideals of technological progress, economic growth and the autonomous subject as providing us with the means to solve ecological problems, ecocentrism conceives of those ideals as part of the problem rather than part of the solution. That is because these ideals signify a world-view in which the human and natural world are strictly divided, and only humans warrant moral consideration. Furthermore, there seems to be no regard for unequal relationships among species and the ways in which some forms of life are vulnerable to the consequences of human action. Therefore, ecocentrists argue that the environmental crisis is not so much a 'shallow' problem of technical mastery but, first and foremost, a 'deeper' problem of mentality, world-view or even spirituality. Environmental problems reveal the unsound relationship to nature we employ within the dominant practices and institutions of modern society. Rather than simple 'problem-solving' the ecocentrist plea is for a reinterpretation of our relationship to nature, our world-view and incentive to change our basic attitudes towards the natural environment (Snik, 1991; Li, 1996; Matthews, 2001).

In his famous book *Deep Ecology* (1973) Arne Naess, one of the most powerful representatives of ecocentrism, set out an alternative, metaphysical world-view. In contrast to what he labelled 'shallow ecology', his 'deep ecology' perceives reality as fundamentally relational; the identity of each individual form of life is a function of its relationships with others, and as such, one form of life flourishing is dependent on the flourishing of other forms of life. This metaphysical insight led Naess to adopt an ethic of interrelatedness, according to which all forms of life are equally entitled to live and blossom. Thus, Naess calls on our desire to live in harmony with the ecosystem of we inextricably are a part and to live in symbiosis with the other inhabitants of this system. Accordingly, Naess pleas for recognition of the values of complexity and diversity of nature. Those values have to be protected, not only because they enhance the opportunities of other forms of life, but because they constitute the integrity of the ecosystem as a whole. Apart from this 'biospherical egalitarianism', Naess presents an alternative notion of selfhood for the self-understanding of the autonomous subject. Rather than a self that employs the environment as a resource for his own purposes, Naess postulates a self that matures by means of identification with ever wider circles of being. These widening circles do not merely imply human sources of identification – our family, community or humanity as a whole – but include elements of our natural environment as well. We identify ourselves in terms of where we were born, the place where we live, the land where we feel we 'belong', our land, our earth. On this view, not only does self-recognition mark the transition from ego to a social self, but from a social self to an ecological self' (Matthews, 2001, p. 221). For Naess, nature comes to have intrinsic value by virtue of the fact that self-realising persons enlarge their 'selves' in such a way as to encompass nature and become one with it (Hargrove, 1992; Li, 1996; Naess, 1973; Matthews, 2001).

Versions of intrinsic value claims

Against the background of this fundamental controversy, the debate focussed on the underlying claims concerning the intrinsic value of nature. Naess and other ecocentrists accused anthropocentrism of a narrow instrumental valuation of nature and a profound disregard of its intrinsic value. However, many anthropocentrists refused to accept the charges of instrumentalism. In response, they criticised their ecocentric opponents of a lack of analytic rigour. In their view, it is senseless and incoherent to speak of the intrinsic value of nature – at least outside the spheres of religion and metaphysics – since any value is contingent upon an act of valuation, and this act can only be pursued by human beings. Without the presence of human beings, nature would be without meaning and value, so they argue. Because we cannot escape ourselves as evaluators, any value we know of must be (at least partly) extrinsic to nature (cf. Les Brown, 1987; Li, 1996). How should we judge these arguments and claims within environmental discourse?

Obviously, there are different notions of 'intrinsic value' at work, depending on the particular discourse in which the term is used and the purpose for which it is used. A lack of agreement among participants on the central concepts they use precludes a clear discussion and analysis on the value of nature. Therefore, I would like to follow O'Neills typology (1992, 2001), who cleared up a lot of misunderstanding and conceptual confusion by distinguishing four notions of intrinsic value that are current within environmental discourse: (1) intrinsic value as non-instrumental value, (2) intrinsic value as objective value, (3) intrinsic value as non-relational value and (4) intrinsic value as ethical standing. Discussions about these four notions of 'intrinsic value' are partly about arbitrary matters of definition, but as the following survey will reveal, choices of definition do have an epistemological and (meta-) ethical bearing as well. Though I follow O'Neill's typology, I will elaborate on the different notions in my own terms. Simultaneously, I will judge these notions on their adequacy, distinctness and coherence, so that a first selection can be made as to which notion(s) to employ within the context of this inquiry.

(1) *Intrinsic value as non-instrumental value*

Within the discourse of environmental ethics the phrase 'intrinsic value of nature' is generally employed in order to mark the contrast with the *instrumental* value we often attach to our natural environment by virtue of the human resources and economic commodities with which it provides.us. Nature is regarded to have intrinsic value if and insofar as the value of nature is not limited to the instrumental use we make of it. Nature is valued, not (merely) as a means to some further end, but as an end in itself: 'Good for its own sake, as an end, as distinct from good as a means to something else' (O'Neill, 2001, p. 164). As such, the predicate 'intrinsic value' can be used to refer to a natural object, an activity or a particular state, that an agent pursues or aims at. In some practical situations these three addressees of intrinsic value are tied together in such an intimate way, that it is hard, if not impossible, to discriminate between them. Suppose, for instance, that we recognise within ourselves an intrinsic desire to protect the wetlands of the Dutch *Wadden Sea* against pollution and degradation, what precisely then is the nature and object of our care? Perhaps we do want to preserve the wetlands because this piece of nature represents a meaning and value that cannot be expressed in instrumental terms of personal benefit, profit or usefulness. But equally plausible is that we care for the wetlands because we enjoy aspects of this extraordinary landscape during our vacation activities: we like to look for seals and feed the sea-gulls from the ferry or we enjoy the reflecting colours of the water streaming along the sandbanks on our walks across the shallows.The activities through which we enjoy practical intercourse with nature are then the object of intrinsic value. Still another possibility is that we want to protect the wetlands against aggressive forms of mass tourism, exploitation of oil companies, overfishing and mechanical cockle fishing, arising from public indignation or because of a particular vision on the future state of the *Wadden Sea*. However, most plausible is that our sense of the intrinsic value of the wetlands is based on a combination of these conditions.

Unfortunately, the general notion of intrinsic value as non-instrumental value is of little help for our purpose. Though this notion connects closely to our every day use of (intrinsic) value claims in ordinary language, its meaning is insufficiently specific to guide us to the particular addressee. Furthermore, this notion has little discriminating power within environmental discourse, because few philosophers will deny the possibility of an intrinsic valuation of nature in this weak sense. Nearly all are convinced that animal species, plants, landscapes and so on should be protected and appreciated not only because of their usefulness for human purposes, but because of their non-instrumental qualities as well. Besides, the recognition that something is of intrinsic value in this general sense does not at all exclude the possibility that it is of instrumental value as well. These difficulties and flaws do not lead us to abandon this notion of intrinsic value from our vocabulary. However, the meaning and status of intrinsic value claims have to be specified in order to be fruitful in our analysis. References to intrinsic value only have power insofar as they call upon concepts that give more specific reasons and claims about the ways in which we express nature as a sources of value.

(2) *Intrinsic value as objective value*

Within the discourses of epistemology and meta-ethics *intrinsic* value is generally defined in opposition to *extrinsic* value, in the sense that one can only speak of the intrinsic value of nature, if and insofar as the value of nature is established independent of human valuation. To argue that our natural world is of intrinsic worth in this strong sense is to make a meta-ethical or epistemological claim: the claim that the value of nature should not be located in the subject or its valuing activity but in the natural object itself. This is to assert a form of realism or objectivism about values. An objectivist stance like this would imply that value is 'found' or 'discovered' in nature rather than 'attached' by human beings. The most influential, contemporary proponents of such a strong view on the intrinsic value of nature place themselves within the Aristotelian tradition of teleological thinking, in which nature is seen as inherently purposeful. All living things are seen as being inhabited by a 'good' striving towards their own realisation. Plants, animals and organisms carry within themselves their own means of flourishing in accordance with an inherent purposiveness or *telos*. Respecting the intrinsic value of nature is to respect the immanent claim of the natural good towards its own realisation. Jonas, for instance, perceives the actual striving of an organism for survival as an articulation of the fact that this being wants itself. Thus, he derives an 'ought' from an 'is' by pointing at the underlying teleological self-affirmation of being. There is an ought-to-be, beyond the is (Jonas, 1984; cf. O'Neill, 1992, 2001).

In a way, this meta-ethical notion of intrinsic value encompasses the former notion of intrinsic value as non-instrumental value. After all, the instrumental categories of means and ends necessarily lose their significance within a teleological framework: it is senseless to ask whether the seed is a means to produce a tree or the tree a means to produce the seed. As Arendt points out: 'Unlike the products of human hands, which must be realised step by step and for which the fabrication process is entirely distinct from the existence of the fabricated thing itself, the natural thing's existence is not separate but is somehow identical with the process through which it comes into being: the seed contains and, in a certain sense, already is the tree, and the tree stops being if the process of growth through which it came into existence stops' (Arendt, 1958b, p. 152).

Strong claims of objective value, like Jonas' teleological claim, insist that it is nature itself – as an objective entity outside of us – that determines its value. Whether or not human beings are willing or able to recognise this value makes no difference at all. Neither whether or not humans relate to nature in a particular way. These extreme forms of value objectivism appear to be at odds with the main ambition of environmental ethics, because it invokes a strict dualism between subject and object. Accordingly, it reifies a dualism between man and nature that environmental philosophers have tried to overcome. As such, radical objectivism fails to appreciate our sensual and sense-making experience of being connected with all surrounding forms of life in a way that constitutes who we are. Argued from a more general epistemological perspective, one could add that extreme forms of value objectivism in a specific sense tend to ignore human freedom and plurality. If the correctness of a judgement about nature is assumed to be judged in terms of its correspondence to 'real nature' as the external object of judgement (acknowledging that full correspondence is unattainable and, thus, a perfect check is unattainable for humans), then there seem to be no grounds for respecting plurality of judgement about natural beauty, value and truth among human beings, apart from the fallibilistic reservation that 'we could be mistaken'. As such, the 'monistic bias' of radical objectivism precludes intrinsic respect for plurality (cf. Lijmbach et al., 2002).

Apart from these strong claims of objective value, there are moderate claims maintaining that the intrinsic value of nature originates from nature while at the same time recognising that our evaluative response to nature co-constitutes the intrinsic value we experience and express. Some authors in the field of environmental ethics take their cue from Heidegger's understanding of human language as a creative response to a call. In this view, we 'draw the word out of silence', the silence that emerges in our encounter with something that is beyond our grasp: the mysterious gaze of a cow, the rhythmic sound of a woodpecker or the breathtaking panorama from the top of a hill. Others depart from a slightly more empirical understanding of our evaluative vocabulary as expression of a *sensus communis*, a common sense: the normative concepts we use to express the 'beauty', 'ugliness', 'truth' or 'cruelty' of nature are understood as the inescapable responses of humans under certain objective conditions. It is through this common language or vocabulary that we share a common world (O'Neill, 2001). In the next sections I will elaborate on these particular insights, since they make it possible to hold on to the intuition that it is not completely up to us how we would like to value nature, as well as to the intuition that it is *us* who express this value by virtue of the fact that nature is valuable *to us*.

(3) *Intrinsic value as non-relational value*

Sometimes, intrinsic value is taken to refer to the value an object has solely by virtue of its intrinsic properties, that is, its non-relational properties. If we appreciate a field of grass, for instance, because of its bright green colours or its fresh smell after a spring shower, then the appreciated value is intrinsic to the field of grass itself. In this particular sense, *intrinsic* value is contrasted with *extrinsic* value as well, albeit in a slightly different sense: here, extrinsic value indicates the value that something has by virtue of a particular relationship with another thing. Suppose, for instance that we value a field of grass because it reminds us of our happy childhood days, because it will feed our cattle for a few days, or because it will serve as a pleasant spot for a picnic, then the value is extrinsic to the object of valuation. This particular use of intrinsic value was first proposed by the philosopher G.E. Moore: 'To say a kind of value is "intrinsic" means merely that the question whether a thing possesses it, depends solely on the intrinsic nature of the thing in question' (Moore, 1922, p. 260; cited from: O'Neill, 2001, p. 165).

If we would choose to follow Moore's definition of intrinsic value as non-relational value, then strictly speaking, many predicates we might use to underline that particular species or natural spots are worthy our preservation, like 'exceptionality', 'scarcity', 'biodiversity', 'energy-richness', 'usefulness', 'nutritional value' or 'level of cuddliness' cannot count as expressions of intrinsic properties, because the question whether a particular animal is exceptional, scarce, energy-rich or cuddly cannot be answered without referring to properties of other animals, species or the larger ecosystem. Accordingly, adherents of this notion would deny the status of intrinsic value to endangered species on grounds of their 'endangeredness' alone. The property of endangeredness can only be acknowledged if an external reality – that is the larger ecosystem and food chain – are taken into account. But how are we then to argue for or against the conservation of natural species or spots if their intrinsic value is the premium criterion? If we want to protect the habitat of the godwit (*grutto*) in order to save them from extinction, should we dwell on something like the intrinsic essence of this bird rather than its place in our landscape, in our farming traditions and our life? Apparently, this use of the intrinsic value claim forces us to separate the natural entity we want to protect from its environment that renders it meaning, beauty or value. Such an act of separation is incompatible with an ecological ethic of interrelatedness and holism.

Furthermore, this notion of intrinsic value fails to appreciate the human stories, rituals and practices in which animals, plants and landscapes emerge as valuable. It is against this background that Weston is opposed to intrinsic value claims and argues for an alternative understanding of *values-as-part-of-patterns*: things are only valuable to us in their appearance within our daily patterns of behaviour, rituals and customs. In his eyes, the acknowledgement of value as constituted by an interconnected 'ecology of values' is obstructed by this rhetoric of intrinisicallity (Weston, 1992). In short, the notion of intrinsic value as non-relational value seems to ignore the fact that value is – by its very nature – the expression of a relationship rather than a quality of a single object or entity: 'Value is the quality a thing can never possess in privacy but acquires automatically the moment it appears in public' (Arendt, 1958b, p. 164). But I am getting ahead of myself as I will come to my analysis later.

Apart from this ethical problem, there are conceptual problems tied to this notion of intrinsic value within the fourfold typology presented here. One might wonder whether this notion of intrinsic value is conceptually clear enough, and if so, whether it is sufficiently distinct from the other notions. In my view, intrinsic value as non-relational value either coincides with intrinsic value as non-instrumental value, or with intrinsic value as objective value, depending on the position one chooses on a meta-ethical level. Those who take an objectivist or realist stance (and pursue a notion of intrinsic value as non-relational value) will regard the valuable properties they find in animals, plants, things and landscapes as intrinsic to these natural 'entities' themselves. Rolston, for instance argues that there is 'intrinsic objective value, valued *by me*, but *for* what it is *in itself*. Value attaches to a non-subjective form of life, but is nevertheless owned by a biological individual, a thing in itself' (Rolston III, 1982, p. 146). Consequently, the value of a particular animal, plant, thing or landscape is conceived as a value it has independent of the value of other things and, more important, independent of human valuation. Thus we arrive at an objectivist claim of intrinsic value: intrinsic value as objective value.

Those who take a subjectivist stance (and pursue a notion of intrinsic value as non-relational value) will argue the other way around: they will talk of 'the valueing agent assigning value to objects solely by virtue of their intrinsic natures' (O'Neill, 2001, p. 965). According to subjectivists, it is possible to speak of 'the intrinsic value of nature' if I attribute value to an object on grounds of properties that I perceive or experience as intrinsic to this object. Properties that borrow their value from my valuation of another object, need or purpose are to be regarded as 'relational' properties and therefore 'extrinsic'. At face value, this notion of intrinsic value might appear plausible, but when it comes down to application in practice, profound problems will come to the fore. Suppose, for instance, that a cattle farmer wants to examine the grounds of his appreciation of Frisian black-white dairy cows, then his search for valuable properties that are evidently *relational* will probably result in a list like the following: a cow fed on concentrates gives about twenty litres milk a day; a mature cow will yield about x euro on the cattle market; healthy cows provide us with juicy meat and so on.

While this list contains doubtful cases, it is even harder, if not impossible, to give an account of *non-relational* properties that are beyond reasonable doubt. Nevertheless, a dedicated farmer might come up with the following properties that underly his sincere care for his cows: the marvellous pattern of black splotches on a white hide that lends every single cow a sense of being unique; the loyal gaze in the cow's glassy eyes; the immaculately pink udders, the penetrating sound of her mooing, the undisturbed grazing to fill her four stomachs and so on. Regardless of whether this answer convinces us, the problem is that we cannot judge the relational or non-relational quality of these properties. Is her gaze an independent property of a cow, or is the gaze, at least partly an answer or mirror to our gaze? Is it possible to judge the farmer's appreciation of his cow's grazing independent of his economic interests in their milk production? And is his appreciation of the mooing not a function of this sound being a sign that they need to be milked? In sum, this notion of intrinsic value is everything but clear and hard to maintain. Insofar as the distinction between relational and non-relational values is clear, it coincides largely with the distinction between instrumental and non-instrumental values.

(4) *Intrinsic value as ethical standing*

Finally, intrinsic value claims are used in a classical Kantian sense. In his second formulation of the categorical imperative, Kant claims that one should respect every human being as a person of intrinsic worth. That is, one should never treat other persons merely as means to an end, but always as ends in themselves as well. Within the field of environmental ethics this Kantian notion of intrinsic value is used to indicate that particular natural entities have 'ethical standing', that they are moral subjects and, *qualitate qua*, belong to our moral community, in which we treat one another as ends in themselves. To say that an animal or plant is intrinsically valuable in this sense is to say that their good must be considered. To claim that a pig is of intrinsic worth, is to claim their right to be a pig and their entitlement to be treated like a pig: for example the right to retain their tails – which are cut off in standard factory farming – or the right to have room to move around and wallow in the mud, like pigs do when they are not in human hands. Among those who pursue these claims, there is, however, disagreement on what kind of characteristic someone or something should have in order to be recognised as a moral subject with ethical standing. What property should count as criterion for the assignment of intrinsic value? Peter Singer's ethical theory is well known in this respect, who assigns intrinsic value to higher animals by virtue of their capacity to suffer pain (Singer, 1993). For liberal contractualists, on the other hand, it is the ownership of interests or the ability to pursue a 'conception of the good' that constitutes moral subjectivity (Feinberg, 1974a).

But we do not need to enter into this discussion, because the relevance of this notion of intrinsic value as ethical standing is primarily limited to the discourse of animal rights and animal welfare. Though very interesting in itself, for the purpose of my discussion this discourse is of secondary importance. The alternative understanding of environmental responsibility that we are looking for embraces the special responsibility for our involvement with animal life, but its scope is much broader. Admittedly, our general attitude towards the natural environment is partly determined by our recognition of animals as moral subjects. But even if we agree with Peter Singer, and acknowledge that 'higher animal species' should be able to count on our care and respectful treatment, this leaves our involvement with the rest of nature – unconscious nature – untouched. So, instead of elevating our acquaintance with this exclusive class of higher animals to the status of being exemplary for environmental responsibility, I would like to take a more broad ecological concern as my framework of reference. Apart from this, in the next chapter, our reactive attitudes towards animals will be examined more closely in order to shed new light on our general concern toward the environment.

Towards an intersubjective claim of intrinsic value

After this preliminary survey we are left with two claims of intrinsic value that seem to be rather coherent and at home within our discourse on environmental responsibility and education. At the same time, these claims represent two contradictory uses: an extremely weak claim and an extremely strong claim. First, there is the weak claim that the value of nature transcends the instrumental use we make of its resources. Within our ordinary use of language the intrinsic value of nature is understood in this weak and general sense as non-instrumental value. Nature is regarded as 'good for its own sake'. Second, there is the more specific, meta-ethical claim that the value of nature is intrinsic insofar as it is established independent of human valuation: intrinsic value as objective value. Henceforth, I will refer to this claim as the strong claim of intrinsic value.

Few philosophers will deny the possibility of an intrinsic valuation of nature in the weak sense. Nearly all are convinced that animal species, plants, landscapes and so on should be protected and appreciated not only because of their usefulness for human purposes, but because of their non-instrumental qualities as well. However, since modern subjectivism is strong within analytic ethics, there has been a tendency to deny the possibility of an intrinsic valuation of nature in the strong sense. The subjectivists start their objection by stressing that every value is contingent upon a human activity of valuation. Callicot for instance argues that: 'There can be no value apart from an evaluator (...) all value is as it were in the eye of the beholder. The value that is attributed to the ecosystem, therefore, is humanly dependent or at least dependent upon some variety of morally and aesthetically sensitive consciousness' (Callicot, 1989, p. 27). In a similar spirit Vincent states: 'The environment becomes worthy through human valuing (...) There are certain necessary conditions for any society to exist and flourish. In short, these necessary conditions derive value only in so far as they provide the conditions for the well-being of human agency. If a healthy and clean environment is a condition (necessary or sufficient) for society and thus human agency, it acquires a derivative value from human agency' (Vincent, 1998, pp. 447–448).

So, in this view we cannot escape extrinsic value. Therefore, beyond the domains of metaphysics and ontology, expressions of intrinsic natural value should be regarded as a *contradictio in terminis* (Brown, 1987, p. 49; Li, 1996). In my view, however, we can – or even should – speak of the intrinsic value of nature in both senses. The weak and strong concepts of intrinsic value are in fact two sides of the same picture of human valuation (though they point to different levels). As I will illustrate in the remaining part of this chapter, *a close study of the use of the weak concept of intrinsic value in ordinary language leads to recognition of the intrinsic value of nature in the strong sense.*

If we take the weak sense of intrinsic value (as non-instrumental value) as a starting point for our analysis, then the question arises: what exactly constitutes the difference between instrumental value and intrinsic value? One might argue that, in the end, all human valuation is of an instrumental kind. Take for instance the beautiful landscape that inspires an artist to paint a watercolour. Though, we would not hesitate to ascribe an intrinsic value to this landscape on the basis of its aesthetic qualities, one might argue that the artist 'uses' the landscape as a 'means' or 'instrument' for pursuing her own artistic end. The same goes for my Sunday afternoon walk in a nearby forest. One might argue that the value I attach to this forest is nothing more than an instrumental value: walking in the forest satisfies my aesthetic or contemplative 'needs'. When an aesthetic value judgement is converted into instrumental terms, the person having the aesthetic experience is depicted as using natural scenery as a trigger for feelings of pleasure (Hargrove, 1992, p. 197). Obviously, these examples point to the limits of instrumental valuation; speaking about aesthetic, existential or spiritual values in terms of needs, ends or interests is highly counterintuitive. In our experience, there is a major difference between consumer needs or economic needs on the one hand, and 'needs' of imagination, contemplation, inspiration, consolation and spirituality on the other. Apparently, a radically different attitude is implied in these ways of relating to our natural environment.

Some philosophers refer to instrumentality in such a broad way that any relation between a person and something or someone outside that person is given an instrumental bearing. Thus, human action and judgement cannot escape the category of instrumentality and the possibility of an intrinsic appreciation of the natural environment is ruled out from the outset (cf. Li, 1996; Brown, 1987). In my view, this tendency to reduce all human values to instrumental values is not only reductionist but ultimately unfruitful, since it would render meaningless many clarifying distinctions within the field of ethics. Not only would the distinction between intrinsic and extrinsic value lose its meaning, but along with it, the distinctions between means and ends, constitutive and regulative ideals, categorical and hypothetical imperatives and so on. My fundamental objection against this reductionist form of instrumentalism is that an untenable picture of the moral person is presupposed like the picture of the 'rational egoist', presented by rational choice theory. A moral person is presented as someone who continuously looks for the most efficient means in order to satisfy their predetermined needs and preferences. As such, one is expected to act in the line of least resistance that provides one with the maximum yield. Obviously, this picture abstracts the moral person from their environment, community, past and future. As we will see in the next section, the understanding of human valuation to which this picture gives rise is ultimately flawed. Rather than satisfying *a priori* needs and preferences, we express and find what we 'prefer' and 'need' in dialogue with significant others and the world around us. So, to say that any valuation of nature we know of is necessarily *human* valuation is one thing. But stating that all human valuation of nature is in the interest of humans is quite another. In line with my criticism, Musschenga is ready to acknowledge that in a sense all human valuation of nature serves human purposes or personal preferences, but he adds that those preferences or purposes are not necessarily of an instrumental kind. Instead of speaking of intrinsic value, Musschenga therefore chooses to speak of the *inherent* value of nature. However, introducing such a new term does not solve our fundamental problem, concerning the epistemological difference between instrumental and non-instrumental value. If one chooses to hold on to the intuition that all value of nature is a function of its service to human purposes, as Musschenga does, it is not clear how one can simultaneously perceive human evaluators as those who appreciate nature on grounds beyond our instrumental relationship (Musschenga, 1991; cf. Koelega, 1995).

In contrast with instrumental valuation, Burms and De Dijn (1995) argue that the recognition of intrinsic value rests on an experience of transcendence. Intrinsic value emanates from our involvement in meaningful 'wholes', larger realities that exceed the limits of our contingent existence and valuation. This existential feeling of interconnectedness, of being part of a significant reality we did not choose gives meaning to our world and ourselves[2]. Experiences like these necessarily imply an awareness of negation; inevitably, something eludes our picture of nature, its value is always beyond our grasp, as if nature resists each human appropriation in order to value, to categorise, or utilise. In other words, that which brings about an experience of meaning, beauty, wonder, imagination or inspiration transcends the experience itself. The awareness of an elusive 'surplus' is often articulated as the experience of a call; we experience 'a call' from those things that go beyond our limited existence: a beautiful layer of clouds, a frightening storm or the amazing flight of an eagle apparently appeal to our feeling for meaning and beauty (Burms and De Dijn, 1995, p. 8).

That the value of nature transcends the valuation of the subject or collectivity follows logically from the fact that not all valuation of nature can be experienced nor represented as the result of an instrumental act – in such a way that the value of nature is determined by our preferences or standards. Ergo, valuable nature transcends valued nature. In fact, our recognition of intrinsic natural value originates from the experience of this 'surplus'. A call emanates from that which eludes our standards and preferences. A more precise description of the nature of this experience will be given as we discuss Kant's category of the sublime and Heidegger's understanding of language as a response to a call.

By recognising that intrinsic valuation in the weak sense (i.e. non-instrumental valuation) rests upon a transcendent experience of a call, we have unintentionally moved from a weak to a rather strong concept of intrinsic value. This being acknowledged, the subjectivist might insist that 'natural value' in this sense still results from *our* subjective responses; individuals will respond to this call by applying their own personal value to nature. However, contra the subjectivist, I would like to stress that it is not *my* response or *your* response, but *our* response that is intrinsically meaningful. As we can delineate from the insights of Wittgenstein, nature can only be significant or meaningful to us within a shared language: 'The limits of my language are the limits of my world' (Wittgenstein, 1923, 5.62). (This does not mean that there is nothing outside of our language, but it does mean that the 'things out there' – the sheer existence of things and our reactive responses that are there like our life – have to be expressed in order to be significant, in order to have a meaning and place in our world). There is no such thing as a private language, in which we express our most inner and individual impressions of nature, without appealing to a *sensus communis*, a common sense appreciation of nature. Every expression of value solicits an understanding in agreement. In other words, we find ourselves constantly negotiating for a personal appreciation of nature with an audience imagined as a tribunal whose understanding, if not whose approval, is necessary for our experience of intrinsic value (Altieri, 1994; Blake et al., 2000, p. 136). So, from the outset, the value of nature is part of the intersubjective practices of our language community. Intrinsic value is neither intrinsic to the subject – the evaluator – nor to the object of valuation – nature itself – but rooted in ordinary language.

Summarising, we may say that our common use of weak intrinsic value claims only makes sense if we accept (some version of) the strong intrinsic value claim. After all, if we recognise that it is possible to reciprocate with our natural environment in a way that cannot be expressed in purely instrumental terms, then we acknowledge the realisation of a transcendent experience. This experience should be characterised as 'transcendent' in a double sense: first, the experience of nature transcends any expression of natural value. Second, any expression of natural value transcends the preferences of the beholder/evaluator. That is, the expression of value is not under voluntary control of the 'beholder', since this expression only acquires meaning within an intersubjective practice, governed by rules of meaningful behaviour that are beyond subjective control. Thus understood, this double transcendence is continuously being mediated by evocative language: not a language representing an object in front of subjects, but a language creating meaning among at the interface of intersubjective experiences of nature. Thus, the value of nature is caught in the openness and movement of language. By virtue of this flexibility and openness, evocative language is able to capture natural value in a way that does not fix nature as an object nor fasten the evaluator to one expressive response. It is in these terms of double transcendence that nature is experienced as an inexhaustible source of value. In other words, the value of nature is not determined by the preferences of the beholder, nor dictated by our intersubjective language rules and standards. To claim that natural value is independent of our valuing in this double sense is to assert a (version of the) strong claim of intrinsic value.

Whereas human beings as individuals are not necessarily at the centre of this valuation, their presence is required and – even more important – their possession of language is fundamental to the very possibility of valuation. Nature should not be defined in opposition to 'culture' – in terms of what is left untouched by human beings. On the contrary, natural entities such as plants, animals, rocks and landscapes are meaningful only if they are somehow fostered in our human practices and, as such, inserted in our horizons of sensibility and significance. In this sense, nature is a dimension of reality, intrinsically part of our human world, rather than a separable entity. Even if we speak of 'wilderness' or 'pristine nature', we refer to those images of 'untouched' nature as we have seen on *Discovery Channel*, on our safari-tour in a zoo, or we appeal to fantasies evoked by reading *Robinson Crusoe* or *The Jungle Book*. However, the value of nature can never be attributed completely to the intentions of the valuing subject. In my view, that is precisely what Michael Bonnett means by stating that 'we are not the *author* of things but the *occasioner* of things' (Bonnett, 2003, p. 687; cf. Rolston, 1982).

In the spirit of Kant's aesthetic judgement of nature

The intersubjective account of intrinsic value as previously sketched requires specification on several points. First, a more detailed understanding should be given of the so-called transcendent experience: what does this experience imply, where does it come from, what or who does it transcend, and how does this experience manifest itself in every day life? These questions will be dealt with in the next section (on the nature of transcendence), since we should first deal with a second issue, concerning the precise nature of the accepted intersubjectivity. To say that the intrinsic value of nature has an intersubjective bearing within ordinary language and everyday practices is not to say that the meaning and value of nature are determined by language in the sense that *subjects* have no say in what they holds meaningful, true or beautiful whatsoever. Neither does the primacy of the intersubjective imply that the *object* of valuation – nature – is a mere construction. Therefore, questions will need to be raised about the epistemological status of subjective and objective claims within an intersubjective understanding of natural value.

In line with my study of civil responsibility in the previous chapter, we will start from the philosophical framework provided by the Kantian tradition of thought on aesthetic judgement. Just as political judgement can be understood analogous to aesthetic judgement, the valuation of nature should be understood in these fundamental terms. Essentially, our appreciation of nature should be conceived of as a matter of taste, as Kant argued in his *Critique of the Power of Judgement* (1790). Taste, not in any trivial sense, but again, in terms of that which constitutes our common sense or *sensus communis*: the power to judge the beauty of something. It is, however, curious to note that the power of taste is commonly considered to be a highly individual sense, too subjective to do the required foundational work (Bonnett, 2003, p. 671). To say that something is 'a matter of taste' is to say that we are not in the position to question the judgement of others since there can be no dispute about one's taste. However, if Kant and his interpreters are right about the nature of aesthetic judgement, then there is more to say about beauty than that it is 'in the eye of the beholder'. Moreover, there is something like a 'beholder's share': a general standpoint, intimately tied to the *sensus communis* that makes communicable judgements of taste possible. By inserting our judgements in a common world of action and speech we anticipate recognition of others, and as such, we appeal to intersubjectively experienced qualities of taste and beauty. Ergo, in our expression of aesthetic judgement we transcend our subjective situatedness as space- and time-specific beings (Kant, 1790; cf. Fischer, 2001).

For Kant, nature as such counts as an object of aesthetic judgement *par excellence*. In fact, Kant argued for the priority of natural beauty over and above artistic beauty, first and foremost, because of the disinterested pleasure we find in nature, consequent upon the 'free play' of imaginative faculties and feelings that the experience of natural beauty arouses (*das freie Spiel der Gemütskräfte*). This is in contrast to human works of art that are more or less designed and intended to elicit feelings of beauty and imagination. Thus, the aesthetic delight we find in nature reveals the aesthetic judgement in its most authentic, i.e. non-intellectual and disinterested form. For Kant, beautiful is that which brings along a disinterested feeling of pleasure (Gadamer, 1960, p. 56–58).

As a *reflecting* judgement the faculty of taste is opposed to the *determining* judgement – normative within Kant's practical reason – not only in that it trades on our imaginative powers rather than our rational faculties, but foremost by virtue of its excessive particularity. The determining judgement subsumes the particular under a given universal rule, principle or law, whereas the reflecting judgement starts from the given particular object for which the universal is yet to be found. In this way, we perceive the object in the unrepresentable fullness of its properties; we attend to the object in its individual particularity. For instance, in the intuition of a beautiful flower, it is a matter of 'sustaining the operation of the powers of cognition without ulterior intention. We linger in the contemplation of the beautiful because this contemplation reinforces and reproduces itself' (Kant, 1790, p. 68). Rather than an activity that consists of applying principles, concepts or rules, judging in aesthetic matters implies a sensual art of attentiveness to the object as given in our experience, unmediated by categories or concepts. As such 'aesthetic pleasure is a pleasure in the *perception of the unreduced presence* of an object and its surroundings, in the instant of whose perception the life of the subject enacts itself' (Seel, 2000)[3]. Thus understood, an experience of beauty is unprecedented. Our desire for the particular object cannot result from a previous desire for the object nor a somehow 'filled' expectation. Such a desire for the existence of the object would bias the judgement and move it from the realm of aesthetics, within which experiences are to be judged disinterestedly, as experiences in their own right (cf. Fischer, 2001, p. 269). Perhaps, that is why every time we are astonished by a beautiful sunset – in itself almost a pastiche or cliché of natural beauty – it is as though we are enjoying the sunset for the very first time. It is because we are receptive in such a way that we experience the sunset as though it takes us by surprise, time and time again, that we are able to see its beauty. And perhaps, it is because we expect a similar expression of beauty when we glance through our photo album, that our pictures of such a sunset are nearly always disappointing.

Within the domain of aesthetic experience, then, to reflect is not to move from the universally given to the particular, but in reverse direction, to go from and beyond the immediately given to the universal. However, as I have outlined in the section on political judgement, this movement towards the 'universal' is not realised by means of rational induction – because then we would have to assume the operation of a pre-given law – but by means of an enlarged way of thinking – *eine erweiterte Denkungsart* – conceived of as the ability to 'to think in the place of everybody else'. To be more specific, our faculty of judgement consists of a twofold mental operation. First, it is by means of imagination that that we can present to our inner sense that which is absent: a familiar landscape, a terrifying snake or a beautiful sky. Our imaginative powers enable us to picture natural things or natural scenes in such a way that we are immediately seized by its beauty or ugliness, its goodness or badness, its truth or illusive nature. Those sensations are highly subjective and cannot be exhaustively represented in language. Therefore, subsequently, we perform an operation of reflection on that which has previously been presented by our imagination. As indicated in the previous chapter, this reflection comprises an act of taking a multiple perspective and making our reflection communicable to others by foreseeing the common sense we share as a community: a *sensus communis*. In other words, we express the things that bring about an experience of beauty within the evaluative language we share with significant others. Thus, we take the evaluative attitudes of other people concerning our expressions of beauty to our own identifications of beauty. Imagination as well as reflection plays a vital role in this transformation from the particular object to the general judgement. Modern interpreters of Kant, like Hannah Arendt, deliberately speak of a 'general judgement' and 'generalisation' rather than a 'universal judgement' and 'universalisation', presumably, because the latter terms invites unwelcome connotations of a transcendental *a priori*. In my further elaboration on Kant's aesthetics I will follow this translation of universality into generality (Arendt, 1982; cf. Cuypers, 1992).

Martin Seel explains how Kant understands the aesthetic experience inherent in the power of judgement as a liberating experience:

> *(...) in the enactment of aesthetic perception we are in a special way free – free from the compulsions of conceptual recognition, free from the calculation of instrumental action, and free from the conflict between duty and inclination. In the aesthetic state, we are free from the coercion to determine ourselves and the world. This negative freedom, according to Kant, also has its positive side. For in the play of aesthetic perception we are free to experience the determinability of our selves and of the world. Where the real stands before us in a fullness and mutability that cannot be grasped and yet can be affirmed, we experience a space containing possibilities of cognition and action which is always presupposed by all theoretical and practical orientation. Therefore Kant sees in the*

> *experience of the beautiful (and even more in that of the sublime) a bringing to bear of man's highest capacities. The richness of reality admitted in aesthetic contemplation is experienced as the joyous affirmation of its vast determinability through us* (Seel, 2000)[4].

Whereas contemporary philosophers might hesitate when it comes to whether it is still sensible to state that the reality we find by immediate apperception 'is presupposed by all theoretical and practical orientation', Kant is commonly praised for revealing the particular moral significance of aesthetic experience. In fact, he describes the experience of natural beauty as a moral experience; an experience that reminds us of our place in this world and the necessities that sustain our being: 'the starry sky above me and the moral law within me'. Kant goes even further by stating that the inherent purposiveness of all natural life – striving towards its own flourishing – reminds us of our own purpose as final destiny of God's creation (cf. Gadamer, 1990, p. 56). We do not have to go as far as this, to acknowledge that there is an exclusive subclass of natural experiences in which we feel that something larger than ourselves appeals to us, or mirrors our desires. Take for instance the experience of a magnificent landscape that suddenly thrusts itself upon us. Well known are the romantic nature poets of the nineteenth century, like Ralph Waldo Emerson and Henry Thoreau, for their exalted celebrations of the healing impact of such magnificent peak experiences on the human soul. In his essay *Nature* (1836), Emerson dwells on our most authentic involvement with nature as an experience that brings us back into a child-like state of innocence and receptivity:

> *The lover of nature is he whose inward and outward senses are still truly adjusted to each other; who has retained the spirit of infancy even into the era of manhood. His intercourse with heaven and earth, becomes part of his daily food. In the presence of nature, a wild delight runs through the man, in spite of real sorrows. Nature says, – he is my creature, and maugre all his impertinent griefs, he shall be glad with me. Not the sun or the summer alone, but every hour and season yields its tribute of delight; for every hour and change corresponds to and authorizes a different state of the mind, from breathless noon to grimmest midnight. Nature is a setting that fits equally well a comic or a mourning piece. In good health, the air is a cordial of incredible virtue. Crossing a bare common, in snow puddles, at twilight, under a clouded sky, without having in my thoughts any occurrence of special good fortune, I have enjoyed a perfect exhilaration. I am glad to the brink of fear. In the woods too, a man casts off his years, as the snake his slough, and at what period soever of life, is always a child. In the woods, is perpetual youth. Within these plantations of God, a decorum and sanctity reign, a perennial festival is dressed, and the guest sees not how he should tire of them in a thousand years. In the woods, we return to reason and faith* (Emerson, 1883).

To say that we experience 'a call from beyond' is to say that within such an experience of nature something happens that requires our further explanation. In its inescapable presence, nature questions our being. Nature appears to us in an appealing form: it requires a response, not primarily in the sense of an articulate answer, but a response in our attitude and action. For instance, the experience of being immersed in an overwhelming fjordland might release within us a strong feeling of futility or piety, as if our being here did not make the slightest difference whatsoever. In harmony with our immediate emotional responses – feelings of comfort, consolation or anxiety – the experience of such a natural setting that entirely surrounds us, urges a particular mode of appreciation that is not under our voluntarily control: its presence makes us speak more softly, it makes us move slower, act more caringly or change our conduct in a different way. Thus, every moment 'corresponds to and authorises a different state of mind', as Emerson writes. Or to paraphrase Michael Bonnett, we are 'commissioned' by the experience of natural things to do this rather than that. Not only natural experiences of great magnitude, but common-or-garden experiences might elicit similar moral responses as well. Consider for instance how we react to the entreating look of a redbreast in our window frame on a freezing winter day, or imagine our response to the sound of a screeching cat. Natural experiences of this kind have a moral significance in the sense that they structure our world in a way that makes particular responses – expressed in words, actions and attitudes – more appropriate than others (cf. Scruton, 1996; Fischer, 2001, p. 268; Drenthen, 2002, pp. 76–80; Bonnett, 2003).

In his *Critique of the Power of Judgement* (1790) Kant refers to the former, exclusive category of overwhelming natural experiences as the experience of the *sublime*. Whereas 'ordinary' aesthetic judgement deals with matters of taste and pleasure, the experience of the sublime opens up a sense of the absolute. This experience is marked by ambivalence: a strong sense of unease, tied to a sense of intense pleasure. For instance, if we are forced to take shelter from a powerful thunderstorm or bear witness to a volcanic eruption, we are likely to realise our human vulnerability and futility in a kind of uncomfortable, perhaps even painful way, while we are simultaneously overwhelmed by a profound experience of beauty. Therefore, some speak of the sublime as a sense of 'delightful terror' or 'negative lust'. However, this experience cannot (yet) be represented or expressed in ordinary language. Initially, we are faced with a notorious speechlessness. The sublime constitutes a transitional category, indicating the transcendent experience of that which is on or just beyond the boundaries of our language and *ipso facto* our world. Here, nature shows itself as the dimension of reality that challenges each every instrumental appropriation on our part. In this metaphysical sense nature can be understood as the inexhaustible source of meaning that simply escapes our concepts and categories because these are necessarily tied to one-sided human perspectives. As such, the experience of the sublime is intimately connected with our ordinary sense of beauty or taste; nature is beautiful in the ordinary sense, precisely because something always escapes our judgement of taste. It is this dimension of elusiveness and infinity that co-constitutes natural beauty. Obviously, there is more beauty to nature than meets our senses (Hargrove, 1989, p. 88;

De Mul, 2002; Drenthen, 2002; Bonnett, 2003). Nevertheless, as Martin Drenthen points out, there are legitimate reasons to doubt the practical significance of sublime nature:

> *Sublime nature withdraws itself from us, it is inconceivable, and it provokes wonder and a feeling of awe. However, it does not allow us to identify its exact meaning or to construct a system of ethics that could justify our actions. Sublime nature reminds us of the fact that there is something 'out there' that is valuable in a way we cannot control, identify, or possess. This sublime nature, however, is not just convenient. Wild nature can be beautiful and sublime, but can also be discomforting, perhaps distressing sometimes. We cannot have one without the other. We can only experience something of value, when we dare to risk losing it* (Drenthen, 1999, p. 8).

Among those contemporary authors within the field of environmental ethics who follow an aesthetic approach to the issue of natural value there are still major controversies between those who hold a rather subjectivist account of value and those who defend a more objectivist or realist understanding of value. The fundamental disagreement between them is primarily about the nature of aesthetic experience. On the one hand, there has been a tendency within modern aesthetics to identify aesthetic experience as a merit of the subject. The competences and attitudes of the perceiver are regarded as prior to the experience: aesthetic experience is thought to result from the exercise of these dispositions (characterisation of Bonnett, 2003, p. 669). These authors account of aesthetic value as a function of our aesthetic response. They tend to define the aesthetic value of an object as 'the value it possesses by virtue of its capacity to provide aesthetic gratification' or please our aesthetic desires (Fisher, 2001, p. 266). Illustrative is William James' understanding of value and beauty as 'pure gifts of the spectator's mind' (cited in: Rolston, 1982).

Objectivist critics retort, however, that to regard natural value or beauty as completely dependent on the perceiver – determined by the eye of the beholder – is to turn nature into a chimera or private projection. Michael Bonnett for instance chooses to hold on to the idea of nature as a primordial reality that precedes us, and whose essence or ultimate meaning is always beyond our grasp. In line with the ancient Greek understanding of nature as *physis* – or the self-arising – he argues that its origins are independent of us. As such, we cannot get behind it. (Bonnett, 2003). Some critics are led by moral indignation about the human arrogance to which subjectivist accounts of natural beauty attest. Notice, for instance the response of John Laird: 'There is beauty … in sky and cloud and sea, in lillies and in sunsets, in the glow of bracken in autumn and in the enticing greenness of a leafy spring. Nature, indeed, is indefinitely beautiful, and she seems to wear her beauty as she wears colour or sound. Why then should her beauty belong to us rather than to her?' (cited from Rolston, 1982, p. 126).

Obviously, a struggle with intuitions like those articulated by Laird will not bring us any further, since these intuitions are in themselves contradictory. Even the ideas of the later Holmes Rolston, who aims at steering a middle course between value-subjectivism and objectivism, remain ambivalent on the issue of aesthetic experience. His intuitions oscillate between the intuition that 'such experiences happen to us without any liberty to refuse them' on the one hand, and the intuition that 'value-judgements have to be decided' on the other (both cited from: Rolston, 1982, p. 127). In my view the main problem is that both subjectivist and objectivist claims of value and beauty suffer from a *petitio principii*: the legitimacy of the conclusive claims is already assumed by the premises; if one chooses to express the question of aesthetic experience in terms of the subjective dispositions that make possible such an experience, then the conclusion will be stated in subjectivist terms since the question *a priori* assumes that natural value serves a subjective function. If, on the other hand, one chooses to phrase the issue of aesthetic experience in terms where objective properties count as a necessary condition for such an experience, then the conclusive claim is likely to have an objectivist bearing. In both cases, the epistemological framework within which the questions are being raised is biased in favour of a subjectivist or objectivist claim. Against the background of these problematic biases, our *intersubjective* framework might prove to be a more balanced starting point for further examination since it does not privilege the epistemological status of the individual subject or the object in advance.

3.2 ON THE STATUS OF NATURE AND HER EVALUATORS

A philosophical investigation of natural value that starts from the primacy of practical intersubjectivity transcends the kind of dualism that postulates a valuing subject standing over and against the natural world as an independent objective reality that has to be grasped. After all, the primacy of the intersubjective implies that any appreciation of nature takes place within our common language and practices in which we interact with nature such that subjective and objective claims cannot be dealt with separately. Rather, meaning-giving practices themselves constitute the value of nature. Within the field of practical ethics, this intersubjective constitution of value is best worked out by those who defend a Wittgensteinian-expressivist perspective on responsibility, in line with the initial framework offered by Kant's *Critique of Judgement* (1790). However, it is of great importance for my purposes, to hold on to the intuition that there is some reality to the idea of nature as an object of judgement on the one hand, and the idea of a subjective agency on the other. After all, any understanding of ecological responsibility involves the idea that there is *someone* who takes responsibility for *some reality (partly) outside of herself*. Let us therefore now consider the epistemological status of the valuing subject and natural object within such an intersubjective framework of value in more detail.

Transcendent nature as an 'object' of judgement

In discussing the status of the 'natural object' within an intersubjective understanding of value, there seem to be only two possible stances to take. If one assumes nature to be *immanent* to our world and worldly expression of nature, then one implies that natural value is itself constituted by the intersubjective practices within which we participate and have contingently agreed upon. If one assumes nature to be *transcendent* – transcending our world and worldly expression of nature – then one implies that natural value is constituted by nature 'out there' and its meaning has to be 'received' from beyond our language. Obviously, such a simple schism does not cover the complexity of the issue at hand. In this section, I will argue that the distinction between immanence and transcendence does not coincide with the distinction between a worldly-intersubjective constitution of natural value and an extra-worldly constitution of nature, understood in Kantian noumenal terms. My aim is to show that it is legitimately possible to hold on to the primacy of the intersubjective as well as to the idea that nature transcends our intersubjective understanding of nature.

As we have seen, our judgement of natural beauty directs our attention to something that transcends intersubjective matters of taste: the sublime. The experience of the sublime constitutes a transitional category, marking the transcendent experience of that which is on or just beyond the boundaries of our language and *ipso facto* our world. Nietzsche expressed the feeling of ambivalence, arising from the 'taciturnity of nature' in his ode to the sunset glow on the seashore:

> *Yonder lies the ocean, pale and brilliant; it cannot speak. The sky is glistening with its eternal mute evening hues, red, yellow, and green: it cannot speak. The small cliffs and rocks which stretch out into the sea as if each one of them were endeavouring to find the loneliest spot – they too are dumb. Beautiful and awful indeed is this vast silence, which so suddenly overcomes us and makes our heart swell. Alas! what deceit lies in this dumb beauty! How well could it speak, and how evilly, too, if it wished! Its tongue, tied up and fastened, and its face of suffering happiness – all this is but malice, mocking at your sympathy: be it so! I do not feel ashamed to be the plaything of such powers!* (Nietzsche, 1982, p. 423; original work published 1881).

In my view, the transcendent call we experience (in the secondary sense) emanates from a tension between that which is significant and meaningful to us, and that which is not (yet). As a result of this tension, we feel that something is questioning us; we are called to give meaning to nature, to find words for the unspeakable. Now, the objectivist might argue that we are in fact implying an objectivist notion of value; we respond to a call that arises from our contact with nature, and try to find words for the value intrinsic to our natural environment. Though this might sound plausible, in my view, it is not correct. In line with Heidegger's notion of expressive language, I will argue it is not Nature that speaks to us, but indeed, it is *language speaking* ('Die Sprache spricht'). For Heidegger the essence of human language is couched in the concept of *Lichtung* (clearing); in our language we disclose a world in which things appear to us as meaningful for the first time. That is, language does not represent a pre-existent reality. Rather, language creates an expressive space in which the essence of things can appear to us. If we speak for instance about the breathtaking panorama we have enjoyed from the top of a hill, we are not (merely) *bringing to light* this experience – describing the height, the palette of colours, distances, the species of trees we could descry and so on. First and foremost, we will find ourselves *bringing about* an experience of natural beauty, evoking within ourselves and our audience a feeling of admiration, desolation, fear or whatever. This expression is not a matter of self-expression (so not subjectivist either), nor an immediate revelation on the part of the landscape itself, but a creative response to a call emanating from language itself. Charles Taylor underlines this evocative understanding of language in his paper *Heidegger, Language and Ecology*:

> *So language, through its telos, dictates a certain mode of expression, a way of formulating matters which can help to restore thingness. It tells us what to say, dictates the poetic or thinkerly word, as we might put it. (....). This is how I think we have to understand Heidegger's conception of language speaking. It is why Heidegger speaks of our relation to language in terms of a call (Ruf) we are attentive to. 'Die Sterblichen sprechen insofern sie hören'. And he can speak of the call as emanating from a silence (Stille). The silence is where there are not yet (the right) words, but where we are interpellated by entities to disclose them as things. Of course this does not happen before language; it can only happen in its midst. But within language and because of its telos, we are pushed to find unprecedented words, which we draw out of silence. This stillness contrasts with the noisy Gerede in which we fill the world with expressions of our selves and our purposes. These unprecedented words ('sayings' is better but 'word' is pithier) are words of power; we might call them words of retrieval. They constitute authentic thinking and poetry* (Taylor, 1995, p. 124).

So, poetic language enables us to express those things that are beyond words, to speak about the unspeakable. In line with this insight, Taylor writes elsewhere that 'the poet makes us aware of something in nature for which there are as yet no adequate words. The poems are finding the words for us. In this "subtler language" – the term is borrowed from Shelley – something is created and defined as well as manifested' (Taylor, 1991, p. 85).

In other words, whereas the incentive originates from outside language (*the interpellating entities*), the transcendent call (*to disclose them as things*) itself can only be heard in the midst of words. We draw the words out of silence, but the silence Heidegger speaks of is a silence surrounded by language. Apparently, poetic or evocative language makes us responsive to a transcendent reality, in a way that enables us to experience and express the beauty or intrinsic value of nature. The ideas of Heidegger indicate that this value can neither be reduced to a subjective experience of respect, nor to an objective quality or voice that speaks to us. Rather, it is by virtue of our language, and its 'telos', that we disclose a world of things that speak to us and demand our respect[5]. Because of its telos, we cannot deploy this language simply as a means to our own ends and purposes. The words transcend my purposes and by that resist any instrumental appropriation of its meaning and value. Moreover, the words require us to listen and care. They engender an attitude of solicitude.

I believe that this understanding of the relationship between nature and language is broadly in line with the ideas of Michael Bonnett. In his monograph *Retrieving Nature: Education for a Post-Humanist Age* (2003) Bonnett aims to retrieve an understanding of nature that allows us to obtain a view of 'a right relationship with nature'. Thus, Bonnett suggests that we should understand nature as 'a dimension of experience that apprehends the self-arising in the material/spiritual world of which we are part, including the powers that sustain and govern it'. Bonnett defends this concept of nature against those postmodernist and constructivist critics who proclaim the end of nature or discharge any understanding of nature as 'mere description' (Rorty), 'social construction' (Giddens), 'simulacrum' (Baudrillard) or a signifier without referent outside of the text (Derrida). In his defense, Bonnett starts with the most empirical assault on the reality of nature, made by Bill McKibben in his famous book with the provocative title *The End of Nature* (1989) in which he states that we live in a post-natural world; since the influence of human activity on earth is such that there is no single spot on earth that is left unaffected by human action in some way, nature has lost its independence. Even undiscovered pieces of pristine wilderness are human-affected by means of our influence on climate change: 'By changing the weather, we make every spot on earth man-made and artificial. We have deprived nature of its independence, and that is fatal to its meaning. Nature's independence *is* its meaning; without it, there is nothing but us' (McKibben, 1989, p. 58). By involving all nature on this globe in our radius of action we have stripped nature of its 'intrinsic value': 'we can no longer imagine that we are part of something larger than ourselves', so now 'there is nothing but us'.

Though an interesting and provoking statement, Bonnett adequately responds to this overly simple conclusion by pointing out that to say that all nature is affected by humans is one thing, but to say that nature has lost its independence is quite another, since this would imply that nature is completely dependent on our purposes and will. Ergo, for nature to lose its independence it would have to be completely under the voluntary control of humans. Even in their most arrogant dreams, anthropocentric hard-liners would acknowledge that this is not the case: natural processes – even our own digestion – proceed independent of our will. That our manipulation of natural forces is taking increasingly pervasive and world-embracing forms does not imply that these forces are under human control. On the contrary, as the fragility of human health indicates, there is a sense in which we are 'constantly dependent upon, and subject to, natural processes at both micro and macro levels', and it is in this sense that 'nature remains always beyond us' (Bonnett, 2003, p. 595). Bonnett, concludes that, by taking 'causal independence' as the hallmark of nature, McKibben tends to conflate 'nature' with 'wilderness'. He assumes that we can only find 'true nature' at the largest possible distance from our human world. But this is a self-refuting concept of nature, because the paradox is that, as soon as we find untouched nature, somewhere far beyond our human world, it ceases to be nature in the strict sense of McKibben, since us, humans, finding and naming nature has already affected and changed it. In this sense 'nature has always already ended', since 'the world we inhabit is always already one transformed by human practices' (cf. Van Zomeren, 1998, p. 7; Vogel, 2002, p. 23; Bonnett, 2003, pp. 594–595). The empirical concept of nature as wilderness is by definition an empty category.

With these latter conclusions, we move towards Bonnett's confrontation with a different strand of post-naturalism: the post-modern version claiming that nature is a human construction, not simply in physical terms but in psychological and linguistic terms. To be more specific, the post-modern claim implies that there is nothing more to nature than its human construction; it is simply a cultural artefact. Following Bonnett, I take the position of Richard Rorty as a paradigm case. In *Philosophy and the Mirror of Nature* (1980) Rorty challenges the fundamental idea that language represents a pre-given reality, or that our knowledge mirrors a mind-external 'natural' world. Moreover, he refutes the representationalist idea that the truth of language and knowledge can be judged in function of its correspondence with this discours-independent reality. Contrary to representationalism and contrary to correspondence theory, Rorty does not believe that we can give useful content to the notion that the world, by its very nature, rationally constrains choices of vocabulary employed to cope with it. In his view, any vocabulary is optional and mutable, and there is no reason, independent of a particular discourse, for choosing this or that particular vocabulary. Rather, we just have descriptions that reflect the particular discourses from which we move to and fro. Truth about reality is always truth about 'reality-under-a-certain-description' and these descriptions are optional – we could always choose others. To quote Bonnett on Rorty's view of language: 'Thus re-describing ourselves and the world is the most important thing we can do – finding 'more interesting', 'more fruitful' ways of speaking – and self-formation rather than knowledge is the goal of thinking. For Rorty, such 'edification' aims at continuing a conversation rather than discovering truth' (cf. Hood, 1998; Bonnett, 2003, p. 597).

Bonnett responds to this 'assault on nature and the natural' by pointing at 'the arrogant meta(physical)-magicianry inherent in these views' and the 'self-absorbed cast of mind that is preoccupied with the active rather than the responsive – itself, a reflection of the modern metaphysics of mastery':

> *For example, it is interesting the way that these authors speak of the idea of nature as 'constructed', which has connotations of some focused (if tacit) agency at work, thus portraying it as the result of some deliberate or quasi-intentional activity – in a certain sense, an invention. While such agency may lie behind the production of many concepts, it hardly seems to describe the genesis of the majority, and, when applied to what hitherto have been taken as overarching or grounding notions, suggests a kind of arbitrariness and human authorship that would evacuate them of authority (…). Such a view does indeed turn nature (and everything else that is foundational) into a chimera. But why should we be seduced by this account? A terminal cancer sufferer is likely to receive little comfort from the suggestion that this is just an optional – and perhaps not the*

> *most interesting – description of his condition. Does Rorty's account not trade on the authority of the term 'description', while simultaneously undermining it? What are descriptions descriptions of? On what basis are some descriptions properly to be held as inappropriate or inaccurate? Such questions cannot be adequately answered exclusively in terms of sets of (optional) local norms. Shared descriptions require shared criteria, but the logic of the notion 'describe' requires also something external that is being described* (Bonnett, 2003, p. 603).

In my view Bonnett correctly points at the voluntarist tendency inherent in this strand of constructivism; the obvious fact that we can only experience, know and imagine nature from within our human forms of significance does not imply that we can simply employ the concept of nature any way we feel nor ascribe it any sort of random meaning. Nor does it mean that we can determine the meaning of nature for ourselves once and for all through democratic-deliberative exercise. There are constraints on what it makes sense to say, to imagine, to believe, and *hitherto* there have been constraints on the meaning of nature that are beyond our control. As such, Bonnett is right in stating that nature is culturally experienced but not culturally produced. While Rorty might agree with us on this point in general terms, some of his philosophical statements might seem to suggest that our linguistic concepts are purely arbitrary and optional from the standpoint of the individual speaker[6] (Bonnett, 2003, p. 597).

While I agree with Bonnett on the constrained nature of giving meaning within a human form of significance, I do not agree that these constraints are necessarily 'external' – located outside, before or beyond discourse. Indeed, descriptions are always descriptions *of* something, but why should this something be of a 'primordial' nature? Obviously, Bonnett wants to secure the idea that our speaking of nature is preceded by a natural reality that provides us with the standards against which the truth of our statements can be judged: 'Internal to the idea of valuing something, as opposed to simply liking or desiring it, is a recognition of qualities – of, in the case of nature, perhaps, independence, diversity, subtlety, delicacy, integrity, etc. – that are not simply the product of human caprice, but that are somehow inherent in the thing itself. They meet a "standard" that is independent of us in the sense of not simply being for us to decide' (Bonnett, 2003, p. 670). In short, Bonnett holds on to a transcendental understanding of nature in order to prevent the measure of man to become the measure of natural things. However, in my view, we do not need to refer to the 'thing itself', in order to find constraints on the possibility of meaning that are beyond human control (and we cannot as will become clear). There are profound reasons for believing that these constraints come from within language, captured as it were within discourse, as Blake, Smeyers, Smith and Standish express in *Thinking Again. Education After Postmodernism* (1998). In their view, the constraints on the meaning of a concept like the concept of nature are not to be conceived as foundational – ontologically prior to discourse – but as constraints that come along with the use of language within a particular discourse, with the need to share references, to share meanings and language with fellow speakers. Thus, the 'intrinsic meaning of nature' is warranted by the rule-governed practices of our language community within which our involvement with the natural world is being captured, rather than by nature 'out there'. However, to say that the constraints on the meaning of 'nature' are internal to discourse and intrinsically social is not to say that language cannot refer to anything outside of language, and that we might just as well give up the subsequent ideas of nature, reality and truth. Rather, the point is that:

> *(...) our access to reality, which is genuine enough, is never unmediated by language or, therefore, by dialogue. Of course we can refer to things outside our language; but reference itself is an activity internal to language. This point is often misunderstood. There are always those who argue that since our vocabulary encodes the things we refer to, it also constraints what we can possibly refer to. This does not follow. Vocabulary is an enactment of our transactions with the world around us, which are nonetheless undetermined. After all (...) we are not immured in a single language, even less a unique discourse (Blake et al., 1998, pp. 30–31).*
>
> *Thus, what we can think is bounded up by the possibilities of shared language, but not by the possibilities of one particular language. So, it is similarly bounded up by the possibilities of human relationships, but not by the parochial possibilities for our own particular society. What does constrain the possibilities of social relationships, and thus of language, is the physical form of our embodiment. It seems to be this alone – this shared life-world of eating, sleeping, sex, death – that accounts for the primordial and extensive agreement in judgements which makes shared language possible for us and ultimately constrains what we can say. We can extend thought and language in some directions, but not in any direction at all – only those we can make sense of with other human beings. That makes for a difference between sense and nonsense, and a fortiori between true and false* (Blake et al., 1998, p. 32).

These ideas on the embeddedness of discourse indicate that post-foundationalism – as articulated by Rorty as well as Blake, Smeyers, Smith and Standish – does not necessarily imply a 'hermetically sealed internalisation of truth to discourse' as Bonnett suggests. The authors do not at all deny the existence of an external reality and truth, but they question the status of appeals to such an external natural reality or transcendent truth within human discourse. Appeals to reality and truth just do not seem to do the job they are commonly expected to do. They do not tell us what to say, what to think or do; they cannot help us to distinguish between true and false statements about the world and they do not guide us in dealing with the moral and cultural problems of our time. As we have seen, these purposes can only be served by the standards and goods internal to the human practice and discourse in which they arise.

Behind this epistemological point, there is of course the underlying critique of foundationalism, which might affect Bonnett's position as well. By suggesting that certain qualifications within human discourse immediately represent qualities of a true or real world Bonnett risks the giving preference to or privileging of a particular class of representations of human experience as 'primordial' over and above other representations, just as classical rationalism, empiricism and transcendentalism privileged certain forms of knowledge as providing us with exclusive access to an underlying reality or a true – rational, empirical or transcendental – world (cf. Heyting, 2000). The idea of privileged access to such an undistorted natural world operates as an independent arbiter, and easily serves as a ground of justification for hierarchical dualisms between the natural and unnatural, male and female, sound and unsound, rational and irrational, private and public and so on. One does not have to be a postmodernist in order to recognise that these dualisms have provided the most horrible projects of exclusion and exploitation in modern times of a justifying rationale. Obviously, in the end, this is what renders the grand narratives and their privileged representations incredibility (cf. Lyotard, 1984).

However, Bonnett keeps away from dangerous implications of this kind by stressing the intertwining of our practical experience and understanding of nature with the human form of sensibility that makes these experiences and understandings possible. Such an intertwining would not allow for a privileged access to a natural reality beyond discourse, precisely because this reality is not construed as something 'out there' but as a reality that partakes in our discourse and experience as well as that which transcends our discourse in an inextricable way. However, by stating that this form of sensibility should be conceived as a necessary condition for our experience and understanding of nature, Bonnett does seem to opt for a transcendental concept of nature as an underlying normative horizon, albeit one that is not completely transparent and therefore cannot be (fully) identified:

> *(...) nature as both a concept and aspect of experience is a deeply constitutive element in our form of sensibility – meaning by this latter, that through which our awareness, cognitive and emotional, occurs and is made possible. We take our form of sensibility to be a product partly of our physiology, relating to our biological needs and capacities, and partly of our culture, relating to our languages and conceptual schemes, which mediate whatever is physiologically given – and which in turn are taken to be the product of a history of interaction with a world whose features, such as the existence of solid bodies, are not all determined by our will and therefore require our accommodation. There are clear parallels here with Wittgenstein's notion of a 'form of life' (...). The argument, then, is that nature in the underlying sense of the self-arising can be construed as a primordial reality in the sense of being deeply embedded in our historically grown form of sensibility. To exorcise it would be to bring about a change so fundamental as to lie beyond our comprehension, if indeed it were even possible (...). The significant point is that our experience which is necessarily undergone within our form of sensibility is conditioned by a concept of nature that is constitutive of that sensibility* (Bonnett, 2003, p. 610).

Now, the question is how this transcendental understanding of nature as an underlying primordial reality, that is always given and as such present to us, in a certain way relates to the understanding of nature as 'the other' to which Bonnett alludes as well:

> *Here (...) would seem to be a case of valuing an aspect of nature not because it gives the valuer pleasure, but through a direct sense of, and respect for, the integrity of its 'otherness'. Of course, a certain pleasure may be experienced in its otherness, but in such a case it is no the experience of pleasure that lends it value, but recognition of its value that gives the experience of pleasure (Bonnett, 2003, p. 634).*
>
> *In bringing us into initial contact with nature, human-centred motives can heighten our awareness of non-human-centred apexes of the world. Through their modulations of, or resistance to, our activity, they stand out as 'other'. And this can be as true of aspects of our everyday practical equipment as of things that we do not think of in instrumental terms: experience of sheer otherness* (Bonnett, 2003, p. 670).

If we start from Bonnett's classical concept of nature as 'a dimension of experience that apprehends the self-arising in the material/spiritual world of which we are part, including the powers that sustain and govern it' (Bonnett, 2003, p. 684), then there is a sensible way in which the experience of nature can be understood in terms of otherness: as the self-arising, nature for us is defined by its independence and otherness from the human. However, in my view Bonnett underscores the inherent (and perhaps inescapable) tension in his metaphysical framework between the dimension of experience that apprehends and recognises 'what is' (given) on the one hand and the experience of otherness – the absent, the excluded, the constitutive outside – that hides behind our present understanding of nature on the other. Sheer otherness cannot be experienced since our experience inevitably shapes 'the other', and thus, our experience of the other is always a one-sided appropriation that reveals and hides particular dimensions of the other. Obviously, Bonnett's understanding of the other is less radical than, for example, the absolute and irreducible Other of Levinas, in whose face I am defenseless and taken hostage in an absolute responsibility (cf. Standish, 2003). But even if one starts from a more dialectical Hegelian understanding of the human relationship to nature as the 'externalised spirit' in whose mirror humans define themselves, then there is a tension between familiarisation on the one hand and estrangement on the other: inevitably, we picture the other that 'speaks' to us, or mirrors us, while, simultaneously, we feel that something eludes this picture, and look for that which is repressed by this picture. In my view, this fundamental tension is a driving force behind our intrinsic valuation of nature.

This fundamental tension between the familiarity and strangeness of nature is radicalised by Martin Drenthen, who takes an existential ambiguity inherent in Nietzsche's philosophy as a starting point for his study of the paradoxical nature of environmental ethics. Whereas Nietzsche leans heavily on 'the natural' in his ideas on human self-understanding, his perspectivism and critique of morality leads him to value all human interpretation of nature as expression of the will-to-power. As such, each interpretation of nature can be seen as an exclusion of alternative interpretations or a violent restriction of nature's expressiveness, and therefore, as a seizure of power. In this tension Drenthen recognises an ambivalence that is 'characteristic of our times' and our understanding of nature: 'On the one hand nature seems to have lost its status as a solid, unambiguous ground for moral judgements, on the other hand we do not seem to be able to articulate certain moral experiences without referring to a more or less normative concept of nature (Drenthen, 1999, p. 6; cf. Drenthen, 2003). The underlying paradox is, of course, that the concept of nature can only exist at places – in our language, our common practices, institutions, religion, art – where its 'pristineness' is being harmed. As such, the word is conceptually problematic if not untenable in a similar way as the words 'God' and 'death' are: all idle efforts to name the unnameable (Van Zomeren, 1998). Drenthen takes this ambiguity to be the core meaning of 'nature' within a normative context:

> *I conclude that in Nietzsche's normative use of the concept of nature, nature means that which in the end cannot be but at the same time always has to be 'grasped'. The fact that nature does not have a moral measure evokes a meaning of nature that precedes and transcends our moral activity. One could say that in an absolute sense nature is something strange and different. But at the same time the notion of this nature functions as a criterion of human self-criticism, that is: it functions within a human interpretative framework* (Drenthen, 1999, p. 7).

Similar to the advocacy of Bonnett for an attitude of 'letting be', Drenthen argues that this paradoxical knowledge need not paralyse any attempt to formulate an answer to the question as to what constitutes a right attitude towards nature: 'The awareness of the radical otherness of nature can lead to a new attitude of listening and respect for nature and awareness of human finitude' (Drenthen, 1999, p. 7). While this field of tension is insoluble, human beings are situated in such a way as to integrate a sensitivity to otherness within their attitude towards nature, thus both Bonnett and Drenthen seem to imply. But now we are getting ahead of issues that will be discussed in the next section on the status of the valuing subject within an intersubjective framework of natural value, to which we will turn immediately after a brief recapitulation of the main argument in this section.

In our examination of the status of nature as an 'object' of aesthetic judgement, our aim was to hold on to the primacy of the intersubjective constitution of natural value as well as to the idea that nature transcends our intersubjective understanding of nature. Inspired by the Heideggerian understanding of language and the sublime, I have argued that the transcendent call we experience emanates from a tension between that which is significant and meaningful to us within our horizon of significance, and that which is not (yet). As a result of this tension, we feel like something is questioning us; we are called to give meaning to nature, to find words for the unspeakable. Whereas the call meets us from beyond, the transcendent call (*to disclose them as things*) can only be heard in the midst of words. Apparently, poetic or evocative language makes us responsive to a transcendent reality, in a way that enables us to experience and express the beauty or intrinsic value of nature. Thus understood, the correct statement that nature is a human construction does not imply that there is nothing more to nature than its human construction, which is a cultural artefact *tout court* that has no reality beyond human discourse. On the contrary, in a sense, nature is defined by its independence of human discourse, and its independence of human will. That nature only shows up within our horizon of significance does not imply that its meaning is determined by it. The paradox is that nature can only be meaningful to us, human beings, if it is sealed in discourse. So, in a sense, only by doing harm to its 'pristineness', can we can understand nature. As soon as one has taken 'independence of the human will' as central to nature, then every effort to name, interpret, imagine, know or value nature, reveals itself as a seizure of power. But obviously we cannot refrain from naming, interpreting, imagining or knowing nature. The only thing we can do is integrate within our attitude towards nature a listening attitude towards that which escapes our present categories or images.

To conclude this analysis, I suggest that we understand nature in similar (but not exactly the same) terms of Bonnett's as *a dimension of experience that apprehends the self-arising in the form of life of which we are part, understood as the ground or very substratum of what we experience as human existence, independent of the human will*. Obviously, my definition expresses a human-related understanding of nature – since nature only appears to us from within our forms of significance, sensibility and acquaintance informed by human concerns and purposes – but is not human centred in the sense that instrumental human purposes necessarily take the lead. This means that 'human beings are necessary participants in, but not in control of the showing up of beings' (Bonnett, 2000).

Inherent to this understanding of nature as the self-arising are four characteristics of nature (on a metaphysical level). I am highly indebted to Michael Bonnett for my summary of them. First, nature is everywhere, it is *ubiquitous* and for situated beings like us *inescapable*. However, to say that nature is everywhere is not the same as saying that everything is nature. Rather, everything partakes of nature in a particular way and to a certain degree, even a personal computer, thus Bonnett would argue: 'Nature as a dimension is everywhere *to some degree*, that is, it is not *everything*. A plastic flower partakes of nature to a degree (…) Wilderness partakes in nature to a very high degree; a piece of computer code, perhaps not at all – or at least only at several removes. Such a view overcomes any absolute dualism of nature and culture, nature and artefacts, nature and humanity – but not in a way that obliterates important distinctions' (Bonnett, 2003, p. 616). Second, nature is *transcendent*, and for bodily creatures like us *elusive*. This characterisation seems to be at odds with the former characteristic but this is not so: nature is inescapable, but its value and meaning do continuously escape our interpretations and imaginations. Nature hides her own sources, and will therefore always be a mystery to us.

Third, nature is characterised by *continuity*, in time as well as space. This means that all forms of life are connected by means of symbiosis and mutual dependency in their ability to flourish and survive. For situated beings like us, this means that we are located in a continuous state of reciprocity and metabolism with our natural environment. Fourth, nature is marked by *irreversible changeability*. Though natural time proceeds *cyclical* – in recurring cycles of bloom and decay – nature continuously changes in unprecedented ways. As such, every natural state is temporary, but one moment in a flux *Panta rhei*. For timely beings like us this means that 'we cannot step twice into the same river' as Heraclites' aphorism goes. We experience nature as *cyclical* – in the recurring experience of the seasons, day and night-time – but nevertheless *unrepeatable* (cf. Bonnett, 2003, pp. 614–617). Among other things, this means that there is no original state of nature – no zero or starting point of evolution – to which we can refer in our present decisions of conservation as a standard of authenticity. If we were asked why we want to save our moors from forestation, an answer in terms of authenticity alone – such as that moors are more natural than forests in this area – would therefore not be conclusive. Natural evolution implies a continual process of disappearance of some natural states and species in favour of the emergence of others. If we choose to conserve particular natural states or species against the powers that be, then our reasons for conservation cannot be convincingly expressed in terms of an original state or original species. Indeed, perhaps we experience some natural states or species as more 'authentic' or 'natural' than others, but the predicate of authenticity then refers to our feeling of familiarity or attachment: we experience these moors or hummingbirds as belonging to our land. Thus, romantically inclined conservationists tend to cultivate a nineteenth century picture of the countryside as a background for an ideal of natural beauty. Accordingly, environmentalists appeal to our sentiment for particular animal species like seals, whales, panda or koala bears, rather than to dirty vermin or bloodthirsty predators.

In the previous characterisation of nature, the 'objective properties' and 'subjective dimensions of experience' mirror one another: the *ubiquity* of nature is the mirror image of the *inescapability* on the part of the situated subject; the *transcendence* of nature is mirrored by the subjective experience of *elusiveness*; natural *continuity* finds its counterpart in human *interdependence*, and the *irreversibility* of natural processes is reflected in the human experience of *unrepeatability* and uniqueness. Furthermore, while Kant characterised the aesthetic experience in terms of the *disinterestedness* of the subject, the romantic philosophers and poets stressed the presumed objective counterpart: the *innocence* of nature. Whether are not this is romantic idle talk, is not the point. Rather, these 'mirrors' are indicative of the kind of the relationship between nature and human discourse. How we mirror our natural environment and *vice versa* will be explored in more detail in the section about the ideas of Merleau-Ponty on the chiasm between man and world, but now, we turn to the other side of the mirror: the subject and her expression of natural beauty.

Human evaluators as 'subjects' of judgement

Now the status of nature as an object of judgement within an aesthetic account of natural value has been illuminated, we turn to the question of the valuing subject: if we argue from the primacy of the intersubjective constitution of meaning and value, how can we hold on to the idea of a valuing subject or agency that takes responsibility for the natural things she cares for? The former elaboration of aesthetic judgement and transcendent nature might elicit a picture of the speaking subject as a submissive mouthpiece of natural value in the name of the *sensus communis*. This picture is evoked by the manifold suggestions of a passive subject that we have touched upon. If natural beauty befalls us, if it thrusts itself upon us, where do we stand as an agent? If we are engrossed in an aesthetic experience of the sublime or immersed in an ordinary experience of natural beauty, is it possible to resist a judgement of taste? And if the meaning and value of nature are warranted by common practice how does this constitution leave room for personal taste and plurality of taste among people? Obviously, the picture of the subject as a mere conduit, silently passing on the judgements of the *sensus communis* without any personal voice whatsoever, does not do justice to our intuitions. We experience ourselves not merely as passive recipients of natural beauty, handed over to the general taste of our community, but – to a certain extent – as active agents as well, who express a personal feeling of comfort or pleasure. Moreover, we sometimes experience that it is possible to hold a dissenting opinion over and against a majority of people who disagree with us. This indicates that there must be some way in which my judgement of natural beauty can indeed be claimed as *my* judgement, i.e. as a judgement performed by me, rather than a judgement of taste received as a projection of others. If we want to hold on to the idea of a subjective agency that takes responsibility for the natural things she cares for, then we will have to point out how the subject is able to express the beauty she finds in nature under her own responsibility.

In order to answer this fundamental question I will start from the analysis developed by Charles Altieri in *Subjective Agency. A Theory of First-Person Expressivity and its Social Implications* (1994). His expressivist account of personal responsibility is inspired by Wittgenstein's analysis of rule-governed behaviour and meaning as use, as well as the Kantian idea of an enlarged way of thinking, underlying our judgements. Within our practices of responsibility Altieri distinguishes between the perspective of the first person – the individual expressing the 'I' – the second person perspective – those significant others who hold the 'I' responsible for her expressions – and the third person perspective – the cultural grammar of the language community according to which responsibility expression and ascription takes place. Altieri now wants to save the idea of a first person agency from the compelling appeal of the significant others (second person) and the coercive grammar of our language community (third person). In other words, he wonders what it is that enables the subject to counterweight the claims of others and to express herself in a way that co-changes the rules of our vocabulary, albeit in a modest way. In order to understand this resistance, Altieri suggests that we should understand the individual on grounds of her desire to *be* someone, not any random person the others make of her, but to be a particular person, to be or to become 'me' in the eyes of others. The subject's desire for recognition of the person she is or wants to be, structures her expressions of what she cares for and the audience she chooses. As I have written before, every expression of value solicits an understanding in agreement. In other words, we find ourselves constantly negotiating for a personal appreciation of nature with an audience imagined as a tribunal whose understanding, if not whose approval, is necessary for our experience of intrinsic value and meaning (Altieri, 1994; Blake et al., 2000, p. 136).

It should be noticed that an 'expression' or 'articulation of value' is not necessarily of a discursive kind – captured in a proposition of the kind *'X is valuable to me'* – but must be understood in more broad terms. We express what we care for in our daily behaviour and the rituals of our body: our love of nature can be read off our style of gardening and the level of my environmental consciousness can be read off the way I handle kitchen waste or deal with water. By expressing the personal pleasure we find in nature we willingly or unwillingly present a picture of ourselves to others. For example, the person who cannot stop testifying of his tenderness for the horses in the meadow along the road – in words and action – displays himself as a lover of animals. Likewise, the person who tells exciting stories about severe hiking tours she made through desert hills will be recognised (and probably wants to be recognised) as a 'tough adventurer' who is able to cope with the uncontrollable forces of nature. Thus, expression of natural value and self-expression are in fact two sides of the same medal. Moreover, by means of such a double expression we simultaneously assume responsibility for our expression of natural value as well as our self-expression. By expressing our engagement with nature within a common language we make ourselves vulnerable to the reactive responses and claims of others – responses of praise, blame, mockery, approval, rejection or indifference – that appeal to us in a way that cannot leave our judgement completely unaffected. Suppose for instance that the conversation partners of 'the adventurer' get annoyed with his 'tough stories' and express their bore, then the adventurer will probably feel the need to respond to this reaction because it puts himself as a person and his engagement with wild nature into question. Our inclination to continuously evaluate our (intended) expressions through the eyes of (actual and imagined) others does not necessarily prevent us from arriving at disagreement with others. On the contrary, throughout my expressions I might want to confront others, I might want to try out alternative views or distinguish myself from them by taking a radically different stance. The critical point is that self-evaluation and evaluation of one's commitments to natural value takes place in a dialogue with others that goes on in a dialogue with oneself: *a dialogue intérieur*.

To be more precise on this particular point, by expressing what we care for in nature and simultaneously expressing who we are, we render ourselves susceptible to the potential appeal of others to take responsibility for the things we say and do. Throughout this process we anticipate agreement or disagreement, praise or blame, depending on the kind of recognition that sustains our sense of who we are and what we stand for. It is against this background of dialogue that the urge for self-direction emerges. I start to realise who I am and what I want in dialogue with others who mirror my 'self' and alternative modes of being. Throughout this dialogue I want to have some control of the picture that others have of me. Moreover, I reflect on the 'selves' presented to me by others (second person) in terms of the strong evaluative vocabulary that is independent of my subjective preferences (third person): the things that we hold beautiful, ugly, worthy, unworthy, good or bad within our *sensus communis*. These strong evaluations are not primarily about the things we desire but about the qualifications that render my desires worthy of desire. We would feel bad if we would not desire this or that. For instance, we would feel bad if we did not clean up our litter after a picnic in the moors. It is against this background that O'Neill argues for the use of 'thick ethical concepts' in the evaluation of our involvement with nature: concepts like 'cruel', 'kind', 'just' an 'unjust' allow for a richer appraisal of non-human nature than the thin ethical concepts generally applied by environmental philosophers, such as 'right', 'wrong', 'good', 'bad', 'has value' and 'lacks value'. According to O'Neill the main difference is that these thick concepts include descriptive as well as evaluative claims. To say that the practices of industrial farming are 'cruel' is to *describe* practices that involve intentional infliction of animal suffering, while simultaneously recognising that the 'cruelty' of this practice cannot be adequately characterised without reference to particular kinds of human evaluative responses to the world (O'Neill, 2001, pp. 173–174).

As Charles Taylor makes clear, such a strong evaluation is generally accompanied by self-referring emotions like feelings of shame, guilt, dignity and pride. For Taylor too, a person is someone for whom particular questions have arisen concerning the good and whose provisional answers to these questions have a place in her self-understanding (Taylor, 1989). In line with this, we can conclude that the subject is able to offer resistance to the claims of others and the rules of meaning giving practices by virtue of her desire to express herself and the beauty she finds in nature in a way that meets the recognition of others. Or, to put it in different terms: underlying aesthetic judgement of nature is a striving after recognition in harmony with one's feeling of what is pleasant (Altieri, 1994; Smeyers, 1997). To recapitulate we can now state that the intrinsic value of nature should be understood as our expression of a transcendent experience that originates from our practical involvement with nature from within our intersubjective practices. As such, I agree with Bonnett that an intrinsic concern for natural environments ought to be viewed less as a consequence of aesthetic judgement but rather as a condition of it (Bonnett, 2003, p. 672).

3.3 TOWARDS AN ETHIC OF ENVIRONMENTAL RESPONSIBILITY

As I have outlined in the introduction to this chapter, my concern is not primarily with the issue of intrinsic natural value as such, but with the underlying issue of environmental responsibility: how should we understand the responsibility we assume in our practical involvement with the natural environment? By expressing – in words and deeds – what we value in nature, we make ourselves susceptible to claims of environmental responsibility, as the analysis in the previous section has shown. However, the question is what kind of responsibility we are dealing with here, and what kind of ideal of responsibility is most appropriate in the light of the problems that threaten our natural environment. Dominant within educational as well as environmental policy is the notion of personal responsibility in terms of accountability: agents are assumed to be accountable for the things they say and do, measured against the publicly specifiable aims and standards they have subscribed to and upon which they have rationally decided. Some authors in the field of educational philosophy and ethics refer to this notion as an expression of the modernist-enlightened ideal of rational autonomy; others speak of the neo-liberal account of 'autonomous chooser' or a legalist account of responsibility. Though there may be slight conceptual differences among these notions of responsibility, what they have in common is they presuppose there being an active subject that (1) chooses the things she cares for, (2) articulates these aims in publicly specifiable principles, aims or standards in a shared language (3) is prepared to justify her daily choices and actions in terms of these specified aims, (4) evaluates her own choices and actions in these terms – does my action contribute to the realisation of my aims in life? – and (5) thus in striving for self-direction in lif the things she says and does are informed by independent rational judgement rather than imposed by arbitrary circumstances or external powers.

Environmental responsibility as rational autonomy;
a general account

This concept of responsibility is highly reminiscent of the Kantian glorification of the autonomous person: the person that owns the power of self-rule and self-legislation and therefore should have 'the courage to think for himself'. For Kant, a person acts autonomously when she is inclined to judge and act in accordance with a self-chosen law. For this law to be a *moral* law, it should be accepted on universally held, rational grounds. In line with the first formulation of the categorical imperative, one should be able to ensure that the maxim underlying one's will could function simultaneously as a guideline for universal legislation. In line with the second categorical imperative, one is supposed to respect every human being as a person of intrinsic worth. That is, one should never treat other persons merely as means to an end, but always as ends in themselves (cf. Peters, 1998; Dearden, 1998).

It would be hard to exaggerate the importance of the ideal of personal autonomy for the practice and theory of contemporary education. Education seems to appeal to the human striving for 'authorship' or self-direction in life. In particular, liberal philosophers of education value education as a vehicle for the acquisition of knowledge, skills, virtues and attitudes required by this moral imperative. In contemporary educational jargon the ideal takes on various forms, ranging from 'reflective autonomy', 'rational autonomy' and 'moral autonomy' to 'self'-related terms like 'self-regulation', 'self-determination', 'self-government', 'self-discipline', 'self-realisation', 'self-education', 'self-improvement', 'self-control', 'self-restraint', 'self-examination', 'self-justification', 'self-support', 'self activation' and mental abilities such as 'critical thinking' and 'open-mindedness'. This arbitrary list of autonomy-related aims and principles, indicates that liberal education eventually aims at a certain way of life; a life of self-examination, inspired by Socrates' adage that 'an unexamined life is not worth living'. Adult, autonomous people will make sure that the opinions, convictions and judgements they develop are 'their own' and not the result of some kind of indoctrination, coercion, or irrational temptation. Children should be stimulated to become aware of the rules, conventions, powers and habits that structure their lives and subject them to rational reflection and criticism. By doing this, it is argued children will gradually develop their own codes of conduct, their own personal styles of thinking, judging and acting.

The ideal of personal autonomy appeals to the human striving for ownership or self-direction within a life. The autonomous person desires to be the author of his life story. His identity does not coincide with those of his family, political party, religious community, music scene or whatever – but transcends the plurality of social identities ascribed to him. He will make sure that the opinions, convictions and judgements he develops are his own, and not the result of some kind of indoctrination, coercion, irrational temptation or some indiscriminate adoption of customs and traditions which are mistakenly taken for granted. In a similar spirit Dearden wrote about autonomy as 'a new aim in education, which requires that what a person thinks and does in important areas of his life cannot be explained without reference to his own activity of mind' (Dearden, 1998, p. 453). The ideal of personal autonomy is pursued, not merely because society requires adults to be independent, self-supporting and so on. Moreover, liberal education is inspired by the conviction that an autonomous life is in itself worthwhile.

In light of the major technical and economic developments of the previous century, and in tandem with processes of individualisation, secularisation and globalisation, the mainstream modern ideal of personal autonomy has gone through some fundamental transformations. Philosophers, historians, sociologists and writers of varying disciplines have analysed this evolution in similar terms. It is not my purpose to give an exhaustive analysis of its historical pretexts and explanatory schemes. Surely, that would be beyond the scope of this thesis. Moreover, I am interested in the general process of rationalisation, and its impact on the idea and practice of personal autonomy. Broadly speaking, late-modern western society is conceived of as the result of a process of rationalisation. According to Max Weber's analysis of social action, in the course of modernity a traditional value-oriented rationality (*Wertrationalität*) loses ground in favour of a strategic or instrumental rationality (*Zweckrationalität*). With its emphasis on means-end reasoning, strategic rationality appeals to the human desire for coordination and control over both the physical and the social environment. It is, for instance, the guiding principle behind bureaucracy and the increasing division of labour. When speaking of 'the iron cage', Weber refers to the ubiquity and inescapability of this form of rationality. Furthermore, he convincingly argues that the rationalisation of social action leads to a 'demystification of our worldview': a loss of meaning. That is, we tend to value our worldly activities more and more in instrumental terms, rather than as worthwhile in themselves. Students, for instance, might flock to university in order to enhance their career prospects, rather than appreciating the intellectual challenge, or the social and political life on the campus (Ellwell, 1999).

Beyond the loss of intrinsic meaning, this process of rationalisation leads to a loss of personal freedom, as Jürgen Habermas describes in his thesis of the 'colonisation of the life-world'. The open, free and intrinsically meaningful communication, typical for the life–world, is being corrupted and suppressed by the strategic rationality of the system. In line with this analysis, though in quite different terms, Jean-Francois Lyotard reveals how the 'end of the grand narratives' paved the way for an uncompromising performativity: an obsessive stress on efficiency and effectiveness, in order to optimise the performance of the system. Values like efficiency and effectiveness are often taken as instrumental qualities, employable for every possible purpose. Thus it would seem, the moral quality is in the hands of the engineer. However, as Lyotard shows, these qualities are anything but value-free. They are in the service of a performative system, a system which requires any type of meaning to be commensurable with any other meaning and thus subjecting the heterogeneity of language to being subsumed under the regime of a general law, that is, to be reduced and homogenised to fit the logics of techno-economic performativity. As such, this system is hostile towards every meaning, or value that escapes its categories: 'The application of this criterion to all of our games necessarily entails a certain level of terror, whether soft or hard: be operational (that is, commensurable) or disappear' (Lyotard, 1984, p. xxiv).

Of course, the analyses of Weber, Habermas and Lyotard do not converge. In fact, they diverge immensely. However, what their theses have in common is the acknowledgement of a heterogeneity of forms of rationality, one of which threatens to become normative in a hegemonic way: a strategic *casu quo* instrumental rationality as is standard in economic and rational choice theory. Within this type of rationality, the sole activity consists of looking for the most efficient means to given ends. Human activities are not valued in terms of intrinsic worth, but judged by reference to the balance between costs and benefits. For the most part, the ends and benefits are given with the quasi-neutral 'needs', that are prevalent within the existing economic and political structures and power relations.

In fact, this rationality can be seen as a radicalisation of the grounding metaphysics of the Enlightenment in which knowledge becomes a value in itself. Everything we experience is valued as the instance or reflection of a *universal* rationality. It is this appeal to universality that gives our experience the status of justified knowledge. Other forms of rationality or other sources of knowledge – such as our bodily experience of the world around us or our respect for particular others – are subsumed by this craving for universality. The presentday reflection on 'universal laws' takes on the form of an appeal to global principles that operate within a global market economy. The knowledge of our school curriculum, for instance, is judged more and more in light of the requirements of the global knowledge economy. School knowledge ought to contribute to the yield maximisation of markets and the competitiveness of one's own country, as documents on ESD state[7].

The emergence of instrumental rationality has not left our conception of personal autonomy untouched. Broadly speaking, in the course of modernity, a morally substantial conception of personal autonomy has given way to an impoverished one. Whereas autonomous acts and judgements used to be embedded within substantive moral frameworks – usually given with a particular tradition, ideology or religion, guided by universally held principles of consistency and respect for persons – people at present seem to regard personal autonomy more and more as a matter of individual need satisfaction. Obviously, the so-called 'end of the grand narratives' has given rise to a host of individual lifestyles, extremely susceptible to the impulses of consumer society. The autonomous person is seen as a person promoting his own interests and satisfying his own needs. Whether or not these needs or interests should be regarded morally laudable or repugnant is out of question. So, the needs and preferences of the individual are *a priori* declared sovereign: something is good because I like it. And, apparently, I am not an autonomous person if I am not smart enough to get what I like. What is crucial here is that the needs and preferences themselves are not reflected upon, but taken for granted. Therefore, they even acquire a status of authenticity, as if they were the most natural expressions of personal needs. This is the false feeling of personal freedom, so sharply parodied by Lou Reed in his song *I'm So Free*:

> Yes I am Mother Nature's son
> and I'm the only one
> I do what I want
> and I want what I see
> It only happened to me
> I'm so free
> (Lou Reed, *Transformer*, 1972)

The emergence of instrumental rationality has not merely influenced our stance towards nature, but is apparent in educational practice and theory as well. For instance, behind contemporary proposals for educational 'reform' in New Zealand, Jim Marshall found implicit notions of personal freedom and choice, which are clearly inspired by neo-liberal thought and technocratic rationality. Marshall refers to the underlying ideal of the educated person as the notion of 'the autonomous chooser': 'Students (…) are presumed to be persons not merely capable of deliberating upon alternatives, and choosing between alternative educational programs according to individual needs, interests, and the qualities of programs, but it is assumed that it is part of the very nature of being human to both make, and want to make, *continuous* consumer style choices' (Marshall, 1995)[8]. This 'personal freedom' is not self-evident, but contains a particular kind of freedom, and therefore a particular subjection of the self as well. According to Marshall, freedom and subjection in terms of choice are to be regarded as a product of 'busno-power' (referring to Foucault's notion of biopower): the emphasis on the activity of choosing rests on a behaviourist doctrine, adopted by new economic theories. These theories 'see an autonomous chooser as perpetually responsive to the environment. In which case the autonomous chooser is capable of infinite manipulation by the structuring of the environment (…). The logical implication is that one's life becomes an enterprise – the enterprise of the autonomous chooser' (Marshall, 1995)[9].

A similar notion of the educated person is manifest in Dutch educational policy and reform. In general, there seems to be a shift from a substantive ideal of personal autonomy, in terms of self-actualisation and moral self-determination (Langeveld, 1963), to notions of personal autonomy, inspired by a negative conception of freedom. In contrast with the traditional notion of autonomy, as constituted by a particular culture, political and moral community, these notions of autonomy seem to opt for an emancipation of the individual in the absence of any other person, moral framework or community. This person comes close to Musil's 'man without qualities'; a person without history and social context. The individual is thrown back upon her own primary preferences, tastes and needs. In short, to be an autonomous person is to calculate what is in one's best interest and then to act accordingly. That is, finding the most advanced and efficient means for satisfying one's needs and preferences. In this perspective, values of social responsibility, environmental awareness, political participation or solidarity are only of minor or derivative importance (cf. Weijers, 2001).

It is not hard to see that, from an ecological perspective, there are many charges against this ideal of personal autonomy, some of which we have touched upon before. First, the ideal gives rise to an instrumental attitude towards nature. Within a framework of instrumental rationality, our natural environment can only be appreciated and protected insofar as it contributes to human interests. Animals, plants, landscapes and so on, are merely regarded as resources for human consumption, production and exploitation, rather than being valued by virtue of their intrinsic qualities, i.e. as sources of meaning, beauty, imagination, awe and wonder. Second, as a result of our independent, instrumental worldview, we threaten to lose our fundamental sense of belonging to something *larger than life*, a sense of being connected to our natural habitat, in a way that makes us vulnerable, and dependent on the ecosystem of which we are integrally a part. According to many environmental philosophers our rational and autonomous lifestyle gives rise to an objectifying distance towards our natural environment, whereas the environmental crisis requires us to define ourselves not in opposition to, but in continuity with the natural world that we share with other animals, plants and inorganic nature (Naess, 1989; Bai, 1998). Thus, ecological responsibility implies a close personal involvement, a sense of interconnectedness and 'attunement'.

Third, the instrumental ideal of personal autonomy rests on the implicit assumption that there are essential human needs – independent of personal preference, or our location in time and place – that have to be satisfied in order to flourish. These needs are regarded as self-evident. Therefore, we do not need to scrutinise them. Consequently, in this view, it is possible to make a sharp distinction between natural and artificial needs. The first category of needs comprises the basic human needs, like the need for food, shelter, safety and so on. The latter needs are seen as not strictly necessary for human flourishing, but artificially created by society; fashions, trends, advertising, popular culture in general and so on. Those are the luxury goods we can do without. From an environmental perspective such a distinction seems to be useful and attractive: it enables us to determine – once and for all – the difference between human 'need' and 'greed'. Although would I definitely agree with Gandhi, who reminded us that 'the earth has enough for everyone's needs, but not for some people's greed', this is neither a value-free statement nor an empirical description of the earth's bearing capacity, but a moral adage. Every statement about human needs involves a moral standard. For, to say that a human being needs food is to say that he will not measure up to an understood standard unless he gets food. In his paper on the nature of needs-statements Dearden concludes that every statement about needs is value-laden and culture-bound: 'it is false to suppose that judgements of value can thus be escaped. Such judgements may be assumed without any awareness that assumptions are being made, but they are not escaped. The deceptively value-free concept of 'need' does more than foster the illusion of being purely empirical, however, for its use often leaves obscure just what the values are that are being assumed, even when the attention is turned to making these' (Peters, 1970, pp. 32–36; cf. Marshall, 1995; Dearden, 1998, p. 258).

More specifically, the Belgian philosophers Burms and De Dijn argue that 'the ideology of needs' – based on the assumption of essential human needs – misconceives the very nature of human desire. According to their analysis, we only desire for X if, and insofar as X means something to us. Our objects of desire are not given beforehand, but they are created by means of imagination, identification, valuation against the background of the cultural conventions of the *sensus communis*. For instance, unlike (most) animals, who feed themselves – immediately dictated by instincts of survival – human beings cultivate their eating rituals. This is not to deny the biological necessity of food for human survival, but to say that this necessity can never be fully separated from the meaning of our needs. We do not merely desire objects because of their utility or 'consumability', we long for appreciation, recognition and prestige as well. Empirical research has shown that, in time of crisis, people are inclined to cut back on so-called necessary goods, rather than save on luxury articles. Furthermore, since these needs are only meaningful to us within a given horizon of significance, the recognition of needs is culturally and historically situated; they change from time to time, and vary from place to place, even from person to person. Consequently, Burms and De Dijn argue that it is indeed impossible to distinguish between 'need' and 'greed' in a definite way like the neo-liberal ideal of autonomy seems to imply. (Burms and De Dijn, 1995).

The main problem with this talk of needs in empirical terms is that it tends to mask the ideological standards underlying our current patterns of consumption and production. For, there is no such thing as 'a minimal level of welfare', according to which the 'basic needs' of human beings all over the world will be met. Seemingly neutral and universal concepts like 'basis needs' and 'human flourishing' are to a large extend determined by the local, cultural and economic conditions of life. For instance, it is plausible to state that, in times of material prosperity, we are inclined to label a great many more goods as being necessary, than in times of crisis. Likewise, people in third world countries will have a completely different view on what comprises necessity and luxury. A clear-cut indication for this statement can be found in the huge differences between the consumption levels in different parts of the world. In 1995, the high-income countries, home to twenty per cent of the world population, accounted for about sixty per cent of commercial energy use (UNSTAT 1997).

Within the context of this paper, what is crucial, is that the neo-liberal or instrumental ideal of personal autonomy implies an uncritical acceptance of currently felt needs. This *a priori* justification of needs precludes a critical scrutiny of the ideological structures, interests and incentives that renders those needs significance. For instance, according to the 'health experts' who occupy our mass media, the cosmetic industry and the social health standard in general, the need for physical hygiene requires us to shower at least once, but even more appropriately, twice a day, while most of our parents and grandparents managed to maintain healthy lives with one shower or bath a week. It might not be inconceivable for us contemporaries to be incapable of surviving such a physical health regime anymore, since the high level of our hygiene standards has made us extremely susceptible to certain bacteria and other risks to which our grandparents were largely insensitive. I do not want to discuss this issue in detail but restrict myself to noting that something that is presented as a universal biological need is in fact a time-and place-specific social construction. On a local as well as a global level, this 'translation' of needs often serves political purposes. For example, the needs of traditional farmers in third world countries are often framed in terms of intensive expansion – maximising the productivity of agricultural land by increasing the scale of production, intensive artificial fertilisation and the use of genetically manipulated seeds – by international institutions and fund organisations, in contrast to the ideas of the farmers themselves, who want to preserve the good elements of their own traditional farming practices.

Just like environmental philosophers attack the hegemony of economic and technocratic rationality in current society, many philosophers of education argue against the pervasive spirit of instrumental rationality and performativity in contemporary education (cf. Blake et al., 1998; Peters and Marshall, 1996; Taylor, 1992 and more). Some criticism, however, too easily identifies the traditional–liberal conception of personal autonomy with the neo-liberal conception. The development sketched previously, reveals that, though the neo-liberal conception of autonomy can be seen as a recent 'outgrowth' of liberal theory, it should be understood as conceptually distinguishable from the traditional–liberal ideal of personal autonomy. The main differences can now be summarised as follows. First, the neo-liberal conception of autonomy implies an *a priori* acceptance of personal needs. Those needs, tastes or preferences are themselves no object of reflection or critical scrutiny. In fact, they are taken as a natural or neutral starting point for the exercise of personal autonomy. Liberal philosophy of education, on the contrary, is well aware of the fact that every acknowledgement of personal needs reflects a value position. Moreover, the liberal ideal of autonomy requires those needs to be subjected to critical scrutiny (Dearden, 1998, p. 258; Peters, 1970, pp. 32–36; cf. Marshall, 1995).

Second, neo-liberalism and rational choice theory see the individual as completely separable from its community, environment and history. The individual is thrown upon its own private preferences, tastes and needs. How she came to value those preferences, how she developed her tastes or required her needs, is not an issue within neo-liberalism. Most traditional liberals, on the other hand, have given account of the fact that the subjects' moral values are constituted by the traditions and conventions of one's culture and community (this is not to say that individuality can be reduced to social determinations) (cf. Peters, 1998). Moreover, over the last twenty years, liberal philosophers have been forced to respond to the criticism of communitarians. In response to their charges of atomism and individualism, most contemporary liberal philosophers have given credence to the idea that individual morality is embedded or within a social and cultural context.

Environmental responsibility as rational autonomy; a closer conceptual look

Although it is of crucial importance to distinguish between liberal and neo-liberal understanding of rational autonomy, these two are nevertheless intrinsically related to one another. The difference between the liberal and neo-liberal concept of autonomy is only a difference of degree. Due to a common ideological and conceptual history, they seem to be related by means of family-resemblance; the emphasis on social independence, the embeddedness of autonomy in a universally founded rationality, the instrumentally charged vocabulary in terms of individuals, continuously choosing between alternatives, defining their life plan, and pursuing their ends by finding the most adequate means. Whereas the traditional-liberal conception of autonomy did not necessarily lead to a neo-liberal version, it has contributed to such an impoverished conception of autonomy. This happened under the influence of long-term social changes: the emergence of a technological society, the march of the free market, processes of secularisation, individualisation and so on. The liberal ideal of autonomy has proved to be particularly susceptible to determinations and manipulations of an economic nature. In this context Marshall speaks of 'the insertion of the economic into the social' and shows how the individual threatens to become a defenseless 'plaything' of the free market:

> *The logical implication is that one's life becomes an enterprise – the enterprise of the autonomous chooser. But it is not the self of classical liberal theory where the right to formulate one's own purposes and projects was seen as inviolate. It is not just that the insertion of the economic into the social structures the choices of the individual, but that, in a behaviouristic fashion, it manipulates the individual by penetrating the very notion of the self, structuring the individual's choices, and thereby, in so far as one's life is just the individual economic enterprise, the lives of individuals. Needs, interests and growth then become contaminated as both needs and interests become constituted by the insertion of the economic into the social. One's autonomy is penetrated by these economic individualistic needs and interests, setting growth patterns towards, for example, freedom from and choice. If the older liberal version of autonomy had some historical justifications, it is clear that these 'new' autonomous choosers have different needs and interests and that their autonomy is problematic* (Marshall, 1995)[10].

If this analysis is correct, that is, if these two conceptions of autonomy are indeed historically and intrinsically related, then the ecological and educational charges against the neo-liberal ideal of autonomy, at least partly apply to the liberal ideal of autonomy as well.

Although contemporary philosophy (of education) captures a wide variety of conceptualisations of this ideal, on further examination, every concept of autonomy turns out to be comprised of two related, though distinct ideals. That is, to be called an autonomous person is to meet at least two necessary conditions: first, there is the condition of authenticity. The condition of authenticity requires that my desires, beliefs, ideas and choices are actually 'mine' and not enforced, indoctrinated, seduced or coerced by some other person or external conditions. This is the condition of 'ownership'. Second, there is the condition of reflection, demanding of the autonomous person that she subject her desires, beliefs, ideas and choices to rational reflection. Furthermore, she must be willing to act in accordance with her reflective judgement.

Traditional liberal philosophers, like R.S. Peters (1973) and Dearden (1972), tend to focus primarily on the reflective conditions of autonomy. However, when it comes down to the condition of authenticity, they veil themselves in rather vague terms. Peters and Dearden do recognise which kinds of identifications are ruled out by the condition of authenticity – those involving indoctrination, violence, seduction and so on – but are not very clear about what authenticity implies. However, in his paper *Freedom and the Development of the Free Man* (1973), Peters does formulate two conditions of authenticity:

> *Firstly, the individual has to be sensitive to considerations which are to act as principles to back rules – e.g. to the suffering of others. Secondly, he has to be able, by reasoning, to view such considerations as reasons for doing some things rather than others. How individuals develop the required sensitivity is largely a matter of speculation. Obviously, identification with others who already possess it is an operative factor; perhaps, too, a degree of first-hand experience is also necessary – e.g. not shielding young people but encouraging them to take part in practical tasks where there is suffering to be relieved* (Peters, 1998, p. 23).

Peters does recognise that children have to develop 'a sensitivity to considerations which are to act as principles to back rules'. The question remains how children develop this sensitivity, how they come to identify certain goods as intrinsically worthwhile. Perhaps, there is more to be said about this development than Peter's 'speculation' allows us for. If we define authenticity as 'self-identification by means of identifying those intrinsic goods, which make our life worthwhile' (as suggested before) then it is possible to distinguish between two kinds of authenticity or self-identification. According to the illuminating edifying analysis of Stefaan Cuypers, first, there is the traditional liberal conception in which the process of identification is described in terms of choice: we choose the things that really matter to us. Second, there is an alternative conception in which identification is described in terms of care, similar to Socrates' 'care of the soul': we somehow find ourselves in the things we care about.

Underlying the traditional–liberal conception of authenticity is a hierarchical model of the moral person's volitional structure. According to Stefaan Cuypers this model is adequately represented by Dworkin and Frankfurt. Crucial to their understanding of the moral person is the notion of a reflexive will, understood as a reflexive structure of beliefs, desires and other volitions. Frankfurt argues that this structure consists of a particular hierarchy between volitions of a first- and second-order: 'Persons typically not only have desires of the first-order, X desires that p, but also desires of the second-order, X desires or doesn't desire that X desires p' (Cuypers, 1992, p. 6). A standard illustration of this is provided by the person who wants to give up smoking. Most likely, she will still long for a cigarette, but at the same time, she does not want to long for a cigarette anymore. For, she does not wish to be tormented by her own compulsive desire *casu quo* addiction. Obviously, this latter desire is of a second-order, whereas the yearning for a cigarette is of a first-order. Having recognised the difference in moral status, she should be willing to subject her desires of a first-order to the desires of a second-order. If there is conformity between first- and second-order desires, then a person can be said to have acted autonomously. Such an appeal to conformity is often done by environmental spokesmen and educators: we should be willing to make our 'shallow desires' – our desire for a long morning shower, for cheep meat and vegetables, or for frequent holidays by air – subordinate to our desire for a sustainable future. Thus, assuming environmental responsibility is presented as an ability of self-discipline or self-restraint that allows us to resist the temptations of consumer culture.

In Cuypers' view such an hierarchical model of personal autonomy is problematic for two main reasons. In the first place, the exercise of this kind of autonomy leads to a *regressus ad infinitum*, because it is never clear whether or not there is a desire of an even higher order to pursue (Cuypers, 1992, p. 8). How can the environmentalist consumer, for instance, be sure about the higher-order status of the principle of sustainability? It is not inconceivable that this particular kind of environmental awareness is being manipulated by governmental campaigns and commercial lobbies, thereby distracting our attention from the prevailing economic and social structures underlying the environmental crisis. And how can the smoker be sure about the higher-order status of his desire to stop smoking? It is not inconceivable that this wish is being enforced by the aggressive anti-smoking lobby of the media or the social environment. This gives rise to the question, whether there is, or should be a third-order desire, for example, a desire not to give in to social pressure. As long as it remains unclear how we should determine the status of desires, we are forced to appeal to an even higher order of volitions. As a consequence, the perspective of an infinite amount of higher orders opens up. Frankfurt himself acknowledged this difficulty as well. In order to put a stop to the infinite regress, he appeals to the notion of decision or decisive commitment. To decide in this sense, is to promise to oneself: this intrinsic choice is no longer susceptible to external considerations. Consequently, Frankfurt seems to come close to the existentialist doctrine of personal identity, which postulates that a person is the one he chooses to be: 'the person, in making a decision by which he identifies a desire, constitutes himself' (Frankfurt,

1987, p. 170). So, the ultimate self-identification is conceived of as a conscious act of choice; an active and intentional appropriation. As the criticism on existentialism makes clear, such a constitutive act of choosing is hard if not impossible to conceive of or conceptualise.

However, apart from these conceptual intricacies, I agree with Cuypers that there is a more fundamental problem tied to this view of self-identification that reveals itself in its counter-intuitive implications; the notion of self-identification in terms of choice goes against our fundamental intuitions and our aesthetic experience of nature as outlined in the previous section. We experience our identifications and commitments, not merely as things we actively and intentionally seek, but, to a certain extent, as something that befalls us and in we find ourselves: instead of choosing our object of identifications, it chooses us. We experience something like 'a call'. The hierarchical model of self-identification obviously neglects this dimension of passivity and non-intentionality. It would be an illusion to think that ones motives and intentions can be completely transparent to oneself in such a way that we are able to control them. Motivational life has its own dynamics. Therefore, there are limits to the identities we can 'choose'. About these limits, Cuypers writes the following:

> *Moreover, deep identifications seem to elude the conscious image a person has of himself. One can very well decide to become a certain kind of character, and yet, notwithstanding one's act of will, still remain the same sort of character one has always been. A decision to identify oneself only creates an intention to make some desire more truly one's own. If another internal but conflicting desire turns out to be predominant, then one's decision was not whole-hearted. However, this failure of wholeheartedness does not indicate of necessity a lack of will-power or sincerity, since the 'heart' of a person may be located in something that is beyond his conscious and intentional control* (Cuypers, 1992, p. 9).

These objections give rise to a different view on the nature of self-identification. Surprisingly, an alternative conception is given by Frankfurt as well (which has, unfortunately, received less attention). Contrary to the previous notion of self-identification in active terms of deciding, Frankfurt explores an alternative account in terms of 'caring about something'. Crucial to this alternative notion is the insight that, in our daily actions and judgements, we are guided by persons, communities, ideals, projects, stories, paintings, natural spots …, we did not choose, but which we found to be an object of our devotion. We do not choose our objects of care, but we find ourselves caring about an object of intrinsic worth. This process is mainly passive in nature; our identifications escape the conscious intentions we express for having them. Identification in terms of caring is primarily concerned with desires of a first-order. In contrast with the reflective distance, required by the activity of choosing, in this view, identifications spring from a close involvement and contact with the things we care about. Metaphorically speaking, this is an experience of being engrossed in something larger than oneself. One experiences the influence of a strange kind of necessity, a necessity which is not under one's voluntary control: 'He (the moral person – DWP) feels that he cannot help caring so much about this or that as he does. He feels that he cannot bring himself to will otherwise than he does' (Cuypers, 1992, p. 10). Frankfurt refers to this compelling appeal as a 'volitional necessity'. This is not like the compulsive desires of an addict. On the contrary, the notion of a 'volitional necessity' indicates a desire, the moral person is unwilling to resist. In fact, he finds this desire to be constitutive of his identity. The necessity, Frankfurt speaks of, does not feel like the intrusion of an 'alien force', but, on the contrary, as the liberating discovery of something familiar, a part of oneself: 'When they (persons in general – DWP) let themselves be guided in a selfless way by objects which escape their control, they make themselves susceptible to authentic personal liberation. Paradoxically, one has to lose oneself in order to find one's true self' (Cuypers, 1992, p. 11).

It is important to notice that this contrast between self-identification in terms of choice and self-identification in terms of care is presented here as an ideal–typical opposition of concepts. This means that the extreme ends of this opposition do not directly correlate with opposite practices or ideals in everyday life. In practice, the passive and active dimensions of identifications do not necessarily conflict. Often, they perfectly coexist and cooperate within one and the same person. To be more precise about the relationship between the active and passive dimensions of personal responsibility and self-identification, Cuypers argues that the voluntaristic conception of responsibility as choosing *asymmetrically depends* upon the non-voluntaristic conception of responsibility as caring (as restricted by volitional necessity)[11] (Cuypers, 2000, pp. 247–248). That is to say that choosing something in an authentic sense is not opposed to the volitional necessity experienced in our caring for something but instead requires it: 'What prevents identifications through decision making from being self-deceptive, akratic and powerless is that they are restricted and informed by identifications through caring. Those non-voluntaristic identifications make voluntaristic identifications wholehearted' (Cuypers, 2000, p. 248). Herewith, Cuypers underlines Frankfurt's conclusion that: 'Unless a person makes choices within restrictions from which he cannot escape by merely choosing to do so, the notion of self-direction, of autonomy, cannot find a grip' (Frankfurt, 1999, pp. 177–178; also cited by Cuypers). Cuypers' and Frankfurt's argument on this particular point is similar to the argument of Charles Taylor in favour of a horizon of significance, independent of personal choice or preference:

> *It may be important that my life be chosen, as John Stuart Mill asserts in On Liberty, but unless some options are more significant than others, the very idea of self-choice falls into triviality and hence incoherence. Self-choice as ideal only makes sense only because some issues are more significant than others (...). So the ideal of self-choice supposes that there are other issues of significance beyond self-choice (...). The agent seeking significance in life, trying to define him- or herself meaningfully, has to exist in a horizon of important questions. That is what is self-defeating in modes of contemporary culture that concentrate on self-fulfilment in*

> *opposition to the demands of society, or nature, which shut out history and the bonds of solidarity (...). I can define my identity only against the background of things that matter. But to bracket out history, nature, society, the demands of solidarity, everything but what I find in myself, would be to eliminate all candidates for what matters. Only if I exist in a world in which history, or the demands of nature, or the needs of my fellow human beings, or the duties of citizenship, or the call of God or something else of this order matters crucially, can I define an identity for myself that is not trivial. Authenticity is not the enemy of demands that emanate from beyond the self; it supposes such demands* (Taylor, 1992, pp. 39–41).

Autonomous choice can only be meaningful and constitutive for our self-understanding if we experience the particular range of options as worthy our deliberation. Thus, free choice is only possible by virtue of its limited focus; we choose variable options against a background of established truths. We cannot weigh the alternative options presented to us and simultaneously question the background truths that lend these options weight and significance. Thus, the autonomous chooser always stands on the shoulders of unchosen or heteronymously given standards. Within certain limits students are free to choose their subjects and the course their education will take, but this choice distracts attention from the fact that they are not able to choose, for example, which teacher they from which they would like to learn or the classmates with whom they would like to learn. In itself, these latter issues are not likely to be less important to the student nor are they of less importance to the activity of learning. To argue that the promise of freedom inherent to the activity of choosing subjects is limited, is not to say that these other issues should be the subject of choice as well. Rather, it is to argue that the idea of freedom of choice is necessarily limited and does not amount to what education is about. Accordingly, Cuypers argues that: 'Too much opportunity and too many alternatives to choose from corrode a person's self-confidence, paralyse his capacity for decision-making and make him indifferent to what to opt for and choose in the end. Furthermore, if the boundaries of a person's will were themselves among the range of his options to choose from or to decide upon, then this would lose all "substance", (Cuypers, 2000, p. 249).

The notion of a volitional necessity obviously points at the limits of our free or reflexive will. A heavy emphasis on rational reflection, like the traditional ideal of personal autonomy implies, can eventually lead to inauthenticity or self-deception, as Richard Smith argues: 'It will not do to say that people are autonomous to the extent that they give reasons for their actions: as noted above, *rationalisations* have the same structure as good reasons and can be distinguished from them only by our empirical sense of what is and what is not an evasion, a subterfuge, a piece of self-deception' (Smith, 1998b, p. 129). Furthermore, Smith argues that the idea that critical reflection takes us to a level of autonomous evaluation is particularly shaky when the tools of such reflection have manifestly been *fixed*. Imagine for instance – this is my example – the modern high school student that is free to choose whatever subject he likes, but whose choice is carefully regulated and channelled in 'desired' directions by the suggestive option menu and choice technology. In The Netherlands, for instance, it would not cross a student's minds to drop the subject of mathematics, even if her performance and motivation were extremely low. Without mathematics students would simply cut off most learning routes from that point on. Nevertheless, if weak students experience major problems and ask for professional assistance, the response of their teacher might be of the kind: 'Well, maybe you should not have chosen mathematics'. Thus, students are made co-responsible for the 'free' choices that were enforced by the system. Our school system (re)produces a particularly limited kind of freedom that allows students to see themselves as 'autonomous choosers' in a way that enhances the performativity of the system at large: school-effectivity, output-optimisation, status-enhancement and yield-maximisation. The self-as-a-reflecting-chooser has become subject to ideological manipulation. In this light, Smith suggests that we had better trust our first-order desires 'which in their inarticulateness maintain some defense against external violation and manipulation (Smith, 1998b).

If we give too much weight to rational arguments in the process of moral consideration, we might easily become detached from the things we really care about, and consequently, alienated from our 'authentic selves'. In line with this insight, David Cooper argues that there is a stance people often take towards beliefs that are of great importance to them, which could not be regarded as an 'autonomous' stance in the sense being discussed, because it is precisely the refusal to give up a belief despite the judgement that the evidence or reasons go against it:

> *A person will not always be tempted to surrender a religious conviction by his judgement that, of the arguments he has encountered, those which militate against this belief are the stronger. Now it seems to me that we do not do right to try to shake him out of this conviction. To do so successfully could induce a feeling of self-betrayal, of being bullied by people cleverer than himself, a lack of confidence in his right to 'stick to his guns', even a loss of dignity and sense of individuality. It could be argued that, in a sense, rationality is on the side of someone who does remain with his conviction. He may argue, after all, that views which go against the weight of evidence at a given time often turn out to be right; or that, had he listened to other people or read different books, his judgement about the weight of evidence would have been very different; or that the reasons which go against his belief count as good ones only from a certain perspective, which may be a matter of fashion even if he himself is unable to form a credible alternative. In short, he may offer reasons, meta-reasons, for why the reasons which go against his conviction should not clinch matters for him. But this is far from agreeing with Peters's and Dearden's emphasis on basing beliefs upon reasons. The fact that it is through reason that a case is made against settling all beliefs through reasons alone does not alter the fact that the case is made* (Cooper, 1998, pp. 51–52)[12].

If we understand self-identification as a subjective act of choosing, then our valuation of nature appears as an act of projection and appropriation. Natural entities are valuable insofar as they fit the reflective criteria and categories that enable us to choose them as 'useful', 'valuable' or 'worthwhile'. Within this form of identification all values are in a sense instrumental because they are the product of a rational evaluation of things according to standards that are external to them. The listening attitude inherent to our recognition of intrinsic value contrasts strongly with the imperative of active choice that is at the heart of the liberal ideal of personal autonomy. Even more, our solicitude runs the risk of being corrupted and suppressed by this striving for appropriation of value. In order to safeguard our 'authenticity' the ideal of personal autonomy requires us to control all 'external influences' by means of rational reflection. Paradoxically, we thereby seem to seclude ourselves from the very sources of authentic identification, which include, the silence, beauty and inspiration we find in our natural environment. One could say that we must lose ourselves in nature in order to find its intrinsic value.

Now it is clear that Kantian liberal theory mainly neglects this passive and non-intentional nature of (self-) identification, it is not hard to see how this ideal might contribute to an instrumentalisation of our identification of natural value. Liberal theory urges us to subject our intuitive relationship with nature to rational reflection. Thus, the ideal of personal autonomy requires us to distance ourselves from our natural environment, whereas an authentic identification of nature's intrinsic value – at least partly – requires us to be receptive to the transcendent reality and value that thrust itself upon us. I want to argue that, more than our reflective abilities; we need a kind of receptivity in order to distinguish our authentic commitments from those that are imposed on us in an inauthentic way. By placing too much emphasis on personal autonomy in terms of rational reflection and choice, liberal education leaves little room for a transcendent call to be heard. In contrast to the ideal of personal autonomy the educational ideals of care and authenticity imply that sometimes we should give up our striving for autonomy, and lose ourselves to that which speaks to us. Understanding identification and responsibility in terms of care seems to do more justice to the object of identification, to the value inherent or intrinsic to the aesthetic experience of nature. In the next section, I will explore how we should conceive of the 'caring subject' and how this subject can be more receptive to the value embodied in natural experience itself.

Environmental responsibility as care

The previous analysis of the liberal ideal of responsibility as rational autonomy shows that this notion is not primarily false or inadequate but shallow; the ideal fails to acknowledge that in the end any choice on what to value in nature rests on a primordial belief and care for the *Umwelt* that precedes rational evaluation and choice. Without the recognition of this underlying involvement as a caring person, the ideal of rational autonomy gives way to an ultimately formal and anthropocentric notion of environmental responsibility. Furthermore, as we have seen in the previous chapter, this formal understanding of responsibility in terms of accountability for the consequences of our actions in the future does not reach far enough in time since the condition of reciprocity (in the narrow sense as mutuality) does not hold in the relationship between generations. For these reasons, we have to come up with a more substantial and sustainable understanding of environmental responsibility that would amount to our existential involvement with our natural environment.

In the search for a fruitful framework to explore our relationship with nature more closely in terms of the underlying caring involvement, we touch upon Nel Noddings' theory of care as outlined in *Caring. A Feminine Approach to Ethics and Moral Education* (1984). In my view, we should take our cue from this theory for two main reasons. In the first place, Noddings' phenomenological characterisation of the caring relation leaves much room for an aesthetic experience of value by virtue of the fact that the subject of our care – the *cared-for* – is not presented as an objectifiable or otherwise controllable entity that we relate to as an optional other, but as something or someone whose presence addresses us as a whole person and commands our attention in an involuntary way. This understanding of the other's appeal is similar to Frankfurt's understanding of a volitional necessity inherent in our caring for something. According to Noddings it is the caring attitude that lies at the heart of all ethical behaviour. As such, Noddings expresses her existential belief that people do not primarily flourish in their pursuit of independence or self-control but realise themselves within caring relationships. It is in our dependence on those we care for, that our self emerges in an authentic way. Accordingly, I hope to convince the reader that in our involvement with nature we experience an appeal that makes us realise who we are and what we desire for in life.

In the second place, the ideas of Noddings are fruitful because of her distinction between *natural* and *ethical* caring; from the everyday life experience of caring for someone she derives an ethical ideal of caring that has a power of expression and validity beyond the natural caring experience, intimately tied to the present and the particular. Notice ought to be taken of the fact that the predicate 'natural' is tricky here and might cause misunderstandings because of its connotations with an essentialist understanding of human nature, independent of time and place. Such an understanding would obviously go against the intuitions of particularity and otherness that are so fundamental to the experience of caring (which will be our next topic of discussion). Noddings claim is of a more empirical kind; the adjective 'natural' expresses her anthropological assumption that the desire to care and the desire to be cared for are inextricably tied to our situated experience of what it means to be human. Though Noddings is more specific in her analysis of the caring relation, her intuition is similar to Heidegger's notion of care as the human mode of being in the world (*Dasein*), 'covering an attitude of solicitousness towards other living beings, a concern to do things meticulously, the deepest existential longings, fleeting moments of concern and all the burdens and woes that belong to human life. From his perspective we are immersed in care; it is the ultimate reality of life' (Noddings, 1998, p. 40). According to Noddings, the fundamental double desire to care and be cared for originates, not from an innate essence or a natural destiny, but from a shared human condition: our infancy in which we all experience what it means to be physically nursed and cared for. It is within this shared experience of being cared for that we develop fundamental caring intuitions and sentiments towards particular others.

In situations in which we immediately answer our intuitions and subsequently respond to the needs of the other that elicits a caring sentiment within us, we are caring in a natural way. Suppose, for instance, that I find on my garden path a young robin that fell out of his nest. We would probably, make the effort to place the vulnerable bird back into his nest and unite him with his mother without deliberation, even without seeing this as a moral act. We just do it. That means, we immediately and intuitively respond to that which happens to us without an appeal to a moral principle or ideal. Such an appeal would be redundant or idle since our intuitions do the job. Our desire to care coincides with a potential duty to care. Evidently, in situations like this most of us can rely on reactive attitudes that are part of our 'nature'. There is an 'I must' in play here, though not an ethical one, but a necessity in the existential sense in which we respond to a crying baby: 'When we see the other's reality as a possibility for us, we must act to eliminate the intolerable, to reduce the pain, to fill the need, to actualise the dream. When I am in this sort of relationship with another, when the other's reality becomes a possibility for me, I care' (Noddings, 1984, p. 14).

Apart from these natural caring situations, we will experience situations in which we do not desire to care for the needful other but still think we have to respond to the needs of the other. Suppose for instance that it is not a lovely little robin but a young crow that I find on my garden path. Suppose furthermore that I have a low estimation of crows in general, and for crows in my backyard even less, since they chase away the lovely little robins and titmouses and eat the bird food I hang out in the trees for the little birds. At this point, my caring attitude is not at all evident, but perhaps, my moral intuitions do require me to care for this particular bird. Noddings argues therefore, in situations like this one, we appeal to ourselves as caring persons. On the strength of my past caring experiences – my care for the robins and titmouses – I know myself as a caring person, or more particularly, as someone who cares for helpless birds. And it is because I am attached to this self-understanding that I am able to call upon this image of myself as a caring person, *casu quo*, lover of animals and act in accordance with the sense of duty – the ethical *I-must* – that arises from this understanding. This appeal to myself as a caring person will be incited and coloured by personal remembrances of caring experiences in my past that were fulfilling in their tenderness and intimacy. Ergo, it is my ideal self that commands me to care for others if my natural caring responses fall short. Thus, we sometimes experience the influence of a moral imperative that involuntarily forces itself upon us, in situations in which we would not be 'naturally' inclined to care. Our experience of the internal 'I-must' is not decided upon, nor does it follow from a transcendental duty to care, but it arises from an *evolving ethical self*, shaped in congruence with one's remembrance of caring and being cared-for (Noddings, 1984, 1998; Noddings and Slote, 2002). Evidently, we can leave the imperative we experience in such situations unanswered or simply ignore them. Moreover, we can deny or actively resist the imperative, but this would somehow diminish our ideal and, consequently, violate our ideal self. It is in this light that Jan Bransen speaks of the 'best alternative of oneself'. If we care for a dog or a cat we do not primarily *identify ourselves with* the needs or interests of this particular pet, but we *identify ourselves as* persons that care for animals (Noddings, 1984, p. 80; Bransen, 1996).

If we want to succeed in extrapolating Noddings ethical ideal of caring to our relationship with the natural environment, then we should be able to locate the source of environmental caring within (1) our intuitive desire to care for nature, to respond to its appeals and (2) our desire to see ourselves as persons who care for nature, which implies an affirmation of (1). However, the former elaboration on the intersubjective nature of aesthetic experience gives cause to an amendment: (1) and (2) should be completed with a third-person desire, because an ethical ideal or ideal self can only be maintained if it is sustained by social values, images, ideals and expectations concerning care in our society (as manifest with regard to the so-called myth of motherhood): (3) the desire to be recognised as a person who cares for nature by others, not merely by those who appeal to my caring – the *cared-fors* – but third persons as well.

At face value, such an extrapolation of Nodding's ethical ideal of caring in the context of environmental responsibility appears to be promising. However, as soon as we enter into the significant details of Noddings' theory, we find ourselves confronted with fundamental problems, mainly due to the fact that the ethics of care were modelled according to the interhuman caring relation. The particular kind of personal reciprocity that Noddings assumes to be essential in a caring relation is absent in the relationship with our natural environment, or so it seems. And it is this kind of reciprocity that Noddings assumes to be conditional for such a relation to flourish: my desire to care must be received and answered by the *cared-for* in order to sustain my identity as the *one-caring* and the subsequent ideal of caring. But if this picture is correct, what is the exact nature of the required reciprocity? Apparently, there is mutual dependency between the *one-caring* and the *cared-for*, but this interdependence is of a different kind than the contractual reciprocity as mutuality in Rawls' theory of justice. The caring dependency is not grounded in the mutual threat they pose to one another's resources, but an existential dependency: the one needs the other – and *vice versa* – to *be* oneself. The *one-caring* can hardly resist or ignore the appeal of the *cared-for*. One has to respond to the other's needs to maintain one's self-understanding as a caring person. The same goes for the other party: the *cared-for* can only act and speak for herself in her relation with the *one-caring*. The uniqueness of both persons appears within the intimacy and engagement of the caring relation. Thus, being human means being-in-relation.

Noddings characterises the kind of reciprocity peculiar to the caring relation in more detail by defining the involvement of both parties in terms of their personal dispositions. About the *one-caring*, Noddings writes in terms of *engrossment*: a full, that is open and non-selective receptivity to the *cared-for*. In such a receptive state of consciousness we are seized by the need of the other, or as Simone Weil writes: 'The soul empties itself of all its own contents in order to receive into itself the being it is looking, just as he is, in all his truth' (cited from Noddings, 1998, p. 40). More particularly, Noddings described the *one-caring* as marked by *motivational displacement* and *empathy*: 'the sense that our motive energy is flowing towards others and their projects. I receive what the other coveys, and I want to respond in a way that furthers the other's purpose or project' (Noddings, 1998, p. 41). The good of the other is my good. However, Noddings underlines that engrossment and motivational displacement do not tell us what to do; they merely characterise our consciousness when we care.

The other party in the caring relation, the *cared-for*, is in a sense subjected to the care but she is not completely passive. She enhances the vitality of the relation by responding to the care of the *one-caring*, albeit in a minimal sense. Noddings typifies this reactive responsiveness in terms of *reception, recognition* and *response*: 'The cared-for receives the caring and shows that it has been received. This recognition now becomes part of what the carer receives in his or her engrossment, and the caring is completed' (Noddings, 1998, p. 41). Even if this response is little – as in the coos, wriggles, smiles and cuddles of an infant – the contribution of the one cared-for is not negligible. On the contrary, without this contribution the caring relation cannot survive. These responses make care-giving a fulfilling experience. Consider for instance how desperate parents are when their infants do not respond 'normally' to their care. Or imagine how desperate a teacher would be if he does not experience any response from her pupils. Thus, the feedback I receive from the *cared-for* is essential to sustain my identity as the *one-caring* and the subsequent imperative to act in the spirit of my evolving ethical self.

The former description of caring might elicit a dualistic picture of the caring relationship in which the positions of the *one-caring* and the *cared-for* are more or less fixed. This suggestion is false, or merely holds in exclusively asymmetrical relationships like that of a mother and a baby, but mature relationships are generally characterised by mutual care: 'They are made up of strings of encounters in which the parties exchange places; both members are carers and cared-fors as opportunities arise' (Noddings, 1998, p. 41).

In her chapter on our care for animals, plants, things and ideas, Noddings acknowledges the possibility of a genuine reciprocity between human and non-human beings that allows us to speak of a caring relationship. This reciprocity is not merely operative in our sentimental involvement with pets or cuddly animals, but even an industrial farmer may experience every now and then an appeal to care for his animals, emanating from the sensual contact with his cattle. So, even in relations with animals, which appear to be purely functional of nature, an acquaintanceship can emerge within which some degree of mutual understanding is possible and a caring appeal is likely to emerge. However, caring relations with animals largely remain a matter of one-way communication. Humans care for animals, but apart from the popular stories of 'wild children' that were nurtured by wolfs (Kaspar Hauser or Romus and Remulus) and the fictional adventures of 'gifted' animals like *Lassie*, we have no real indications that a similar caring relation in the reverse direction is possible. Our present form of life and common sense do not allow us to recognise animals as *ones-caring* in the full sense the way humans do. Ergo, the inherent reciprocity remains asymmetrical. This asymmetry is not a natural fact or a claim of human exclusiveness. It is indeed possible that our growing sensitivity to animal welfare at present develops in a direction that does allow us to entertain more or less equal caring relations with animals in the future. Perhaps, the relationship of a blind person with its dog comes near to what I envisage, but here, this is a matter of speculation.

Beyond any doubt or speculation is the fact that animals can be sincere *cared-fors*, in the sense that they respond to our caring in a way that helps to sustain ourselves as *ones-caring*. Cats for instance start to purr, nibble, lift their heads and stretch towards the one they are addressing, whereas dogs wag their tails and respond in another species-specific manner. As we get familiar with one particular animal family, we come to recognise its characteristic forms of response and address. Subsequently, we will, perhaps unconsciously, anticipate these animal responses. We start to act in a way that evokes a familiar response of an animal. Thus, our interactions get richer and more complex in such a way, that action repertoires, patterns and rituals are likely to emerge in our common practices. As a consequence, we will gradually feel more sustained as *ones-caring* and more sensitive to the appealing *I-must* that originates from this recognition. Perhaps this is why parents entrust a pet to their children, so that they learn to care.

The question is now whether an ethical ideal of caring can be sustained by such a relationship between humans and animals. In order to test the generalisability of our care for animals, Noddings wonders whether the person that cares for her own cats with all her heart and soul should feel committed to care for the rancid stray cat, appearing at her front door as well. In the end, Noddings answers this question affirmatively; once I engage in intimate interactions with cats, I will grow sensitive to their typically feline appeal when it comes to my caring for them. This sensitivity is of such a nature and intensity that it becomes part of my ideal self: I like to see myself as person that cares for cats. This ideal enables me to care for cats in situations in which my encounter with a cat does not elicit a 'natural' caring response within me. In the absence of such an intuitive inclination to care my appeal to the ethical imperative does the job. Consequently, an ethical *I-must* will emerge from every encounter with a cat that appears at my front door. In this way, my caring sensitivity might naturally extend to even wider circles of animal life. However, my neighbour or any other person who lacks this caring sensitivity to the needs of cats will not experience such an appeal and we therefore cannot blame them if they neglect the needs of our street cats. Furthermore, my care will remain limited to cats, and perhaps some other similar animal families in my immediate neighbourhood, but I will never feel addressed as a *one-caring* by rats or bugs. In this light, Noddings concludes:

> *What we see clearly here is how completely our ethical caring depends upon both our past experience in natural caring and our conscious choice. We have made pets or cats. In doing so, we have established the possibility of appreciative and reciprocal relation. If we feel that the cat has certain rights, it is because we have conferred those rights by establishing the relation. When we take a creature into our home, name it, feed it, lay affectionate hand upon it, we establish a relation that induces*

> *expectations. We will be addressed, and not only by this particular creature but also by others of its kind. It seems obvious that we might live ethically in the world without ever establishing a relation with any animal, but once we have done so, our population of cared-fors is extended. Our ethical domain is complicated and enriched, and to behave uncaringly towards one of its members diminishes it and diminishes us. If we establish an affectionate relation, we are going to feel the 'I must', and then to be honest we must respond to it* (Noddings, 1984, p. 157).

Apart from this exclusive subclass of cases in which individuals have once established a caring relation with particular animals, a general extension of Noddings' ethics of care towards our involvement with non-human nature fails because of a lack of reciprocity between the parties. It is perhaps needless to recall that we failed to justify an ethic of intergenerational responsibility for precisely the same reasons: 'our obligation to summon the caring attitude is limited by the possibility of reciprocity. We are not obliged to act as one-caring if there is no possibility of completion in the other' (Noddings, 1984, p. 149). According to my reconstruction of Noddings' presuppositions, the main problem is not that we do not feel connected to our natural environment. Nor is it impossible to develop a reciprocal relationship with animals, from which a moral appeal emanates. Noddings even insisted on a generalisation of such an appeal to all 'others of its kind'. After all, once we have become receptive to the needs of a particular family of animals, an *I-must* will emanate from every encounter with a strange animal of its kind that appeals to my care. However, the fundamental point is that *no one is committed to enter into an initial relationship* with this or that animal in the first place. The start of such a relation rests on accidentally originating familiarity in past experience or arbitrary choice.

That we do not feel an *a priori* obligation to engage into interaction with cats or dogs is probably due to the anthropological fact that there is no existential need or necessity to do so. We are not, by our very 'nature' or condition as interrelated human beings, in the same relation to animal life as we are in relation to human beings. The primordial relatedness or intersubjectivity Noddings speaks of, is a *human* relatedness and intersubjectivity. Again, this is not to say that animals are not subjects, but that 'we do not have a sense of an animal-as-subject as we do of a human being as subject' (Noddings, 1984, p. 149). This statement leaves open the possibility that animals represent a subjectivity or a way of being-in-the-world, an openness to the world that is not completely open to us (yet). This is why Wittgenstein remarks that 'if a lion could speak, we would not be able to understand him – would not really be able to follow him into his world' (Wittgenstein, 1958, p. 223). We need other humans in our neighbourhood to mirror our existence. We need their company to be recognised as a person. Their lives and responses reflect our desires, ideals and fears. However, our reactive attitudes towards animals are of a completely different – less exclusive – kind, thus Noddings seems to imply. Perhaps it is because we do not engage in the same practices, and subsequently only partly share a common predicament (cf. Burms, 2000). In sum, whereas human relations are premised by an ontological necessity that is universally given with human life on this planet as such, our relation with animals is of a less exclusive *contingent* nature, because 'we might live ethically in the world without ever establishing a relation with any animal'. Thus, we are by our very nature social beings, though we are not necessarily animal lovers. It is against this existential background that we can decide whether or not to engage in a relationship with animals, while we are-in-relation with fellow human beings from the very moment we are thrown into this world.

Elsewhere, Noddings marks the difference between our care for humans and animals as a contrast between a *categorical* versus a *hypothetical* imperative. Noddings obviously refers to her conclusion that our care for other humans is premised by an existential necessity, while our care for animals is contingent upon past experience and personal choice: 'I must if I wish (are am able to) move into relation (Noddings, 1984, p. 86). While this contrast itself is illuminating, I fear that these terms bring with them Kantian connotations of universalisability and consistency that are at odds with Noddings'ambitions. For this reason I choose to speak of a *general* imperative and ideal, rather than a categorical one. As Noddings continues to discuss our relationships to plants, things and ideas, she observes a shading-off from the ethical into the sensitive and aesthetic. Here we are left with personal sentiments, attachments and aesthetic pleasures that are free of obligations. My engrossment in these 'things' might engender an *I-must*, but not in an ethical sense, since our care is not received and answered in a way that sustains my ideal self as one-caring. The self-forgetful pleasure we sometimes find in nature cannot underpin a reciprocal caring relation, thus Noddings. In her view, most if not all of our care for nature is derivative of our care for other persons or is otherwise dependent on interpersonal relations. We care for our garden so that we feel at home in our natural environment. We care for our land, because we eat its fruits to satisfy our basic needs. Perhaps we care for our landscape as well because it is through this land that we feel connected with one another (*heimat*). Nevertheless, if there is an internal *I-must* in connection with our garden or land then it is a mere instrumental must, premised by our needs and wants: we have to care, thus we can eat and drink and sleep and live our lives.

In my view, it is the assumption of a merely contingent reciprocity within our involvement with the natural environment that constitutes the problem here. Moreover, I think this assumption is also false to our intuitions. Perhaps, my relation to trees in general and oaks in particular are of a contingent nature. I can indeed live ethically in a world without oaks or stray cats, but when I am addressed as a person that cares for nature, or an appeal is made to my ecological spirit, then our responsibility is called upon for the involvement with our natural environment *as a whole*. Emerson wrote: 'The whole of nature addresses itself to the whole of man. We are reassured' (cited from: Bonnett, 2003, p. 613). A sense of environmental responsibility does not originate from our interaction with this particular tree, plant or animal, but from the realisation of the particular space we occupy in our natural world and the threads that tie us to this world. Moreover, we look for an answer to the question *who* we are, at least partly in terms of *where* we are: the earthly space we occupy (cf. O'Loughlin, 2002). In a sense, the natural environment is 'a mirror' in which we see, feel, smell, hear and recognise ourselves. We never speak about nature without at the same time speaking about ourselves. With this in mind, I suggest that we search for sources of environmental care in our bodily involvement with the natural environment. After all, it is through our bodily existence and sensual interaction with the things that surround us, that we are connected with the natural life–world, just like it is through language that we are connected with other humans. As such, intercorporeality and intersubjectivity are anthropological facts of life in an equally necessary and similar double way: as gift and task. Like our relation to other humans, our relation to nature is not merely given, but in our involvement with nature we experience an inherent appeal to relate to nature in such a way that it speaks to us. This will become clear as we discuss the fascinating ideas of Merleau-Ponty about the body subject and her world.

Merleau-Ponty on reciprocity as reversibility of perceiving and being perceived

In our perception and bodily involvement with the things that appear in our world we experience a carrying reciprocity: I move towards the thing as much as the things move towards me. In this section I want to study the ethical nature and implications of this reciprocity more closely in dialogue with the ideas of the French existentialist philosopher and phenomenologist of the body and embodied perception, Maurice Merleau-Ponty. In his opus magnum, the *Phenomenology of Perception* (1945) he reveals in great detail how we maintain an intimate relation with the life–world through our corporal being in the world or 'being towards the world' (*être au monde*). This world is not an objective world, stretching out in front of us as an objective reality, but a world that is inhabited and sensually experienced by us. It is through our bodily and sensual existence that we are familiar with the world around us; that we are in the world and intentionally engaged in its affairs. Thus, Merleau-Ponty strives to transcend Cartesian dualism in which the body is valued as an instrument that the human mind employs in order to get a view on the world, to move and act – in order to get along in the world. Moreover, Merleau-Ponty shows that we have a body only by being our body. That is why he chooses to speak of the body as a subject (*corps-sùjet*) that is marked by a pre-reflexive consciousness instead of the body as an object of which we ourselves dispose. Rather than an instrument, our fleshly body is a mode of being, through which we partake of the flesh of the world.

That subjective as well as objective dimensions of being flow together in our sensual experience of the world manifests itself in the ambiguity of our everyday use of the words 'experience' and 'perception'. Sometimes we speak of experience when we are touched by something external, when we experience that something from outside it pervades us. In other situations, on the contrary, we use the words experience and perception to indicate a particular activity or state of ourselves; we refer to perception as if it were a net that we throw out and pull over the world. In his meticulous descriptions and analyses of sensual experiences, Merleau-Ponty shows that these two experiences represent two sides of the same process. Any perceptive experience is *my* experience in which my being is actively involved, but simultaneously, all perceptive experiences are experiences *of something*. Perceptive experience can therefore never be fully understood in terms of passive reception – as classical empiricism wants us to believe – but neither can it be reduced to active human construction – as classical rationalism assumes[13].

Although empiricism and rationalism are diametrically opposed to each other in this respect, Merleau-Ponty accuses both positions for thinking from 'the prejudice of the objective world'; the perceiving subject and the perceptible world are not connected but positioned over and against each other. There is an airtight division between them. As a consequence of this dominant experience of reality as an independent, objective world – material for manipulation and control – we have lost the kind of intimate contact with the life–world that is immediately and intrinsically meaningful to us, even before we our thinking and judging splits this experience in two: the experience of an inside and an outside quality. Following the phenomenological adage of Husserl's to return to the things themselves (*Zurück zu den Sachen selbst*), Merleau-Ponty wants to retrieve the nature of our primordial life–world that precedes knowledge: a pre-reflexive world that cannot be fully known by an independent observer but is embodied and inhabited by it. In fact, Merleau-Ponty points at a primordial relation of meaning that we entertain with the perceptible world, similar to Kants' understanding of aesthetic perception as an attentiveness to the experience of things in their full presence, unmediated by evaluative categories and concepts. But whereas Kant persists in speaking of a 'subject' perceiving an 'object', Merleau-Ponty only speaks in terms of their underlying relation. This relation reflects a pre-given unity between man and world that is not self-chosen and can therefore not be cancelled or cut off by an individual observer. Any promise of such a separation from our world – cultivated by the ideals of self-control, self-management, manipulability and rational choice – so pervasive in contemporary educational and environmental discourse – might operate as a dangerous illusion since it carries within itself an alienation of our natural life-world that sustains our being. Thus, the phenomenology of perception should not merely be read as a meta-ethical criticism of objectivism and subjectivism, but as a form of social criticism as well.

Merleau-Ponty locates the conditions of possibility of human knowledge, reflection, meaning and inspiration within our sensual and bodily mode of existence. Our body thinks, speaks and experiences meaning. It is because the things in our life–world have a similar corporal mode of existence, that man and world are mutually sensitive to one another. This constitutive *intercorporeality*, as Merleau-Ponty labels the fleshly interconnection between perceiver and perceived, is most evident in our tactile sense of touch. In touch, we cover the world with our flesh and return to flesh in order to feel things as real and meaningful. My hand is part of me as much as it is part of the world that it palpates and interrogates. In this respect, Merleau-Ponty wonders how it happens that I give to my hands, in the act of palpating something – say the bark of a tree – in particular, that degree of, rate, and direction of movement that enable me to feel the textures of the sleek and the rough? How is it possible that our bodily senses anticipate a particular sensation that they intentionally reach out into the world? The empiricist would presumably answer that the sensations received by my fingers, together with the sensual stimuli that my visual faculty add to this, form a perceptive representation of the bark, ready to be decoded and transformed into motor impulses by my consciousness. This explanation seems to make sense. After all, the empiricist makes clear how sensual stimuli in general lead to certain motor responses. However, the precise connection between perception and movement is turned into a *black box* by this type of explanation. The empiricist cannot explain why this particular bark attracts my attention while others do not concern me at all. Neither can he make understandable why this particular impression motivates action, while other possibilities of action remain unrealised. The human mind as conceived by emipircism is ultimately empty. It stands indifferently towards the perceived and therefore cannot know what to look for. Why is my forefinger in search of this awkward little knot, of which it instinctively and involuntarily explores the outline? Why does my palpating hand change its direction as soon as it touches the soft moss on the bark? In short, what determines the direction of my perceptive attention and movement?

The rationalist would probably answer that it are not the external stimuli that determine our action, but our mind that grasps aprioristic ideas of all possible forms we find in nature. These mental representations of pure forms accompany all our perceptions. My finger is able to discover a geometric circle in the outline of this knot by virtue of the fact that my mind is already disposed to its intelligible form. According to the rationalist, it is in the dynamic interplay between the transcendental forms of my consciousness and the perceived reality that the direction of my touch is being determined. But obviously, my mind is in charge; my mind employs my tactile senses to explore the properties of this bark by ranging them under general concepts of smoothness, form, temperature, moisture and so on. Apart from more particular problems tied to this position, Merleau-Ponty argues that rationalism also fails to explain the intentional direction of my perceptive attention. Why do my fingers seek particular qualities in the world? Why should this particular knot among all the other irregularities on the bark draw my attention while my mind already knows its form? Here, the problem is opposite to that of the empiricist; the human mind is not empty but too rich. My mind already knows what it is looking for. Therefore, it is unclear why it commands my finger to touch the bark. To find what we expect to find? The intrinsic necessity to touch is absent. If my perceptive engagement were indeed to consist of being on the lookout for the intelligible structure of reality, why would I then my focus on the contingent, why would I pay attention for this particular bark with this awkward oval outline? (Merleau-Ponty, 1962, p. 71).

Merleau-Ponty concludes that empiricism and rationalism fail to understand the contingent direction of perceptive attention because they ignore the intentionality of our body. The searching movement of my finger is not the logical outcome of reflection or the effect of stimulus-activation. Moreover, the exercise of my finger is the immediate answer of my body to the situation in which it finds itself:

> *In the action of the hand which is raised toward an object is contained a reference to the object, not as an object represented, but as that highly specific thing toward which we project ourselves, near which we are, in*

> *anticipation, and which we haunt. Consciousness is being toward the thing through the intermediary of the body. A movement is learned when the body has understood it, that is, when it has incorporated it into its 'world', and to move one's body is to aim at things through it; it is to allow oneself to respond to their call, which is made upon it independently of any representation. Motility, then, is not, as it were, a handmaid of consciousness, transporting the body to that point in space of which we have formed a representation beforehand. In order that we may be able to move our body towards an object, the object must first exist for it, our body must not belong to the realm of the 'in-itself'* (Merleau-Ponty, 1962, pp. 138–139; emphasis is mine).

Consciousness is in the first place not a matter of 'I think' but of 'I can'. We are already interpretative beings on a corporal level on which we are not yet self-reflexive and reflexive towards the world in which we are immersed: 'Bodies are "lived experience". Bodies have understandings of the world which are independent of any sort of cognitive map; these understandings are like a set of invisible but intelligent threads which stream out between the body and that world with which the body is familiar' (O'Loughlin, 1995, p. 3). This lived world or lived space can never be articulated in terms objective time-spatial coordinates. I experience the bark of the tree as 'beautiful', 'tasteful' or 'creepy' before I am able to call the bark 'brown', 'cold' or 'smooth'. Furthermore, we perceive *gestalts* or wholes before we analyse these wholes by means of subsumption under general categories.

We can only feel, see and hear things insofar as we are engrossed by these things. Of course, we can take an objectifying distance to the things, but the created objective world we create by doing this is secondary against the primary life-world in which we are bodily situated. In short, perception is a pre-reflexive act of the body; a body I can hardly call *my* body since I share this fleshly embodiment with other fleshly beings and the tangible world that is flesh. It is this intercorporeality that makes it possible for us to experience the tangible bark without the mediation of perceptive representation or cognitive reflection. My body was familiar with the bark, knew the bark, before I had conscious knowledge of it.

In his last, unfinished book, *Le Visible et l'Invisible* (1964) Merleau-Ponty examines the dynamic relationship and acquaintanceship between perceiver and perceived more closely. Again, he illustrates the nature of this kinship by means of the sense of touch. In the first place, he affirms the idea of intercorporeality by stressing that my hand can feel the bark because they are both of the flesh. Were perceiver and perceived to have a different mode of being-in-the-world, then they would not be able to touch one another. They would not exist in the face of the other, since they partake of completely different worlds. As such, he who touches must not himself be foreign to the world he touches. My body must also be inscribed in the order of being that it discloses to us. I can see because my body is part of the visible world. Accordingly, hearing is a possibility since my body is audible and sonorous. Thus, my body is sensitive to the perceptible world by virtue of the fact that my body itself is perceptible.

Nevertheless, the body interposed is not itself a thing, an interstitial matter, a connective tissue between me and the material world, but sensible for itself. It is for this reason that Merleau-Ponty labels the human body as a *sentant sensible* – a sensible sentient; my body is only sensitive to the world insofar as it is sensitive to itself. Our body raises itself as exemplary reality. It is along the lines of bodily self-experience that our perception extends from our body into the world around us. Thus, our world is being made tangible, visible and audible analogous to our body. In this contact with itself and the world our body transcends its activity. Our body extends beyond itself towards the things in our world, it runs ahead of things and is intentionally engaged with the affairs in the world. The possibility of interpretation and meaning are given with this particular way of being-in-the-world[14]. In the light of this self-transcending activity and corporal being Hans Jonas speaks of a certain 'openness' and 'care' for the world. In a sense, the human being is the measure of all things, not by legislation of human reason, but through its own corporeality. Our contact with reality is not primarily realised by means of cognition and thinking, but through concrete human life that offers resistance to reality and within this experience of personal strength a sense of power and causality in his world develops (Jonas, 1984).

The human body is so important to Merleau-Ponty, because it is the site where the perceptible world turns back upon itself and reveals itself. By means of a reflecting activity the perceptible body comes to perceive itself. Thus, the human body is the initial site where a sense of vision emerges out of the visible, where a sense of touch emerges out of the tangible, where a sense of hearing emerges out of the audible, where a sense of smell emerges out of the olfactory world. Moreover, reciprocity of mutual reflection between perceiver and perceived constitutes the very possibility of perception, thus Merleau-Ponty argues:

> *There is vision, touch, when a certain visible, a certain tangible, turns back upon the whole of the visible, the whole of the tangible, of which it is a part, or when suddenly it finds itself surrounded by them, or when between it and them, and through their commerce, is formed a Visibility, a Tangible in itself, which belong properly neither to the body qua fact nor to the world qua fact – as upon two mirrors facing one another where two indefinite series of images set in one another arise which belong really neither of the two surfaces, since each is only the rejoinder of the other, and which therefore form a couple, a couple more real than either of them. Thus since the seer is caught up in what he sees, it is still himself he sees: there is a fundamental narcissism of all vision. And thus, for the same reason, the vision he exercises, he also undergoes from the things, such that, as many painters have said, I feel myself looked at by the things, my activity is equally passivity – which is the second and more profound sense of the narcissism: not to see in the outside, as the others see it, the contour of a body one inhabits, but especially to be seen by the outside, to exist within it, to emigrate into it, to be seduced, captivated, alienated by the phantom, so that the seer and the visible*

> *reciprocate one another and we no longer know which sees and which is seen. It is this Visibility, this generality of the Sensible in itself, this anonymity innate to Myself that we have previously called flesh, and one knows there is no name in traditional philosophy to designate it* (Merleau-Ponty, 1997, p. 139).

Thus, the human body is the site where a perceiver is born out of the perceived and where an inner space is likely to emerge along with the possibility of meaning and interpretation of an outer space. In fact, it is no longer sensible to speak about the perceiver and the perceived, the inner and outer space, the seer and the visible as two separate entities, since they are inextricably intertwined in such a way that they need the reflection of the other to exist: the seer needs the visible to see itself and the visible glances at the seer to be seen. If we nevertheless choose to hold on to a 'methodological dualism' – which Merleau-Ponty in fact does by continuing to speak of a seer as distinguished from a visible – then the intertwining between the two can best be characterised in terms of a constitutive reversibility i.e. a reversibility of the visible in the seer and the seer in the visible, of the touching in the tangible, of the hearer in the audible and so on. Again, our sense of touch offers the most powerful illustration of such a kind of reversibility. In my hand the touching and tangible meet each other, and because of this encounter, the feeling of the bark is alternately concealed, now in the touching hand, and then in the tangible bark. Thus, I sometimes concentrate on the grip of my hands on the bark, while at another time I concentrate on the delicate way in which the irregular surface of the bark brushes my fingers and invites my hand to move in a particular direction. Analogously, we experience this reversibility in our hearing. My sonorous body resonates in every sound I hear. This is most evident when I hear my own voice. I can hear my voice from the inside – through the vibration and acoustics of my body, or I can attune to the sound of my voice as it resonates and turns back to me through the outer space. Finally, in our vision as well such a kind of reversibility is operative: I can 'palpate' with my look the surface of the bark and explore the seemingly random course of its grooves, but as my look moves along, it easily disappears or loses itself in a particular groove when this captivates me and pulls my look inside. A strange look captivates us, in the way an artist can sometimes feel trapped or looked at by the landscape he is painting. Thus, intense perception can lead my attention away from myself and hide in the things as if it wants to glance back at me from beyond. This is the kind of self-forgetful pleasure we sometimes experience in the sheer thereness of nature, as Iris Murdock expressed earlier (Merleau-Ponty, 1997, p. 139).

In many cases it is hard to tell whether a look is my look, your look, our look or a completely strange look that captivates one. For this reason Merleau-Ponty speaks of a general Visibility or a general Sensibility as an anonymity that is innate to myself as well as the world: it is the Flesh of the world that makes this reversibility possible. However, there can be no complete reversibility between me and the world, as Merleau-Ponty illustrates with the example of the left hand touching the right hand. If my left hand can touch my right hand while it palpates the tangible world, can touch it by touching, can turn its palpation back upon it, the movements of my right hand either resorts to the order of the tangible – but then its grip on the world is interrupted, or it keeps its grip on the world, but then I do not touch my real hand, only its outward surface (cf. Bakker, 1969, p. 45; Merleau-Ponty, 1997, p. 141). Obviously, our sensual experience oscillates in between the two poles of inward and outward experience, without ever reaching the extreme ends. Even if I am completely engrossed by the tangible world, my hands will feel themselves – albeit in a minimal sense – in order to experience what it feels as my feeling. Perception can never be fully ascribed to the perceiver, or to the perceived. Accordingly, I can try to ignore my 'inner voice' and concentrate on my 'outer voice', but I will never be able to hear my voice in a similar way as others hear my voice or as I hear the voices of others. Complete transcendence is not given to the human body. There will always be a tension between the sensible and the sentient body, a tension that cannot be resolved in complete reversibility (Merleau-Ponty, 1997, pp. 139, 141).

Just like a visual look alternately hides in the visible and the seer, so the visible can hide in the palpable and *vice versa*. Due to this reversibility, our senses cooperate within one and the same body. But even more important within the context of education is Merleau-Ponty's insight into the reversibility between my look, your look and the looks of others. My look continuously hides in the looks of others as well as the looks of others can hide in mine. Ergo, the assumed intercorporeality amounts to other seers and thereby includes the primordial intersubjectivity mentioned earlier. If my friend and I enjoy the beauty of a landscape and talk about this or share this experience silently, I do not continuously wonder whether we perceive the same landscape, but act from the certainty of a shared experience. I do not consciously put myself in the place of the other or take the perspective of the other as that of an *alter ego*. Rather, it is through the similar functioning of our bodies and our coexistence within a human form of life that there is common field of vision that unites us as seers with the landscape. It is within this common field of vision that I can see myself. Seeing and being seen – by the landscape as well as by other seers – reciprocate and go together hand in hand. Here, Merleau-Ponty points at the 'chiasm': my receptivity to the others and my receptivity to the landscape mutually include each other and form one horizon or field of vision. Again, Merleau-Ponty alludes to a general visibility that takes possession of us: 'At the joints of the opaque body and the opaque world there is a ray of generality and of light' (Merleau-Ponty, 1962, pp. 422–423; Merleau-Ponty, 1997, p. 142, 146; idem, pp. 508–509).

Finally, of special interest is Merleau-Ponty's philosophy of language, since it connects the intersubjective- expressivist understanding of language (as previously articulated in dialogue with the insights of Wittgenstein, Heidegger and Taylor) with the existentialist phenomenology of the body. Merleau-Ponty reveals the body as a locus of meaning as well as a locus of expression. The same body that gives meaning by dwelling in the flesh of the world, is the same body that expresses within itself what it reaches out for: meaning. My hand, reaching out for something, becomes a gesture, pointing others to the things that occupy me. Likewise, my cry of relief becomes a word, listened to by others as an expression of relief. Thus understood, my palpating body coincides with the expressive body. This means that my expression is not a translation of meaning but rather its very realisation through my bodily searching. Merleau-Ponty underlines that meaning cannot exist within a private language: 'A thought that would be satisfied to exist in and of itself, without the obstacles of words or communication, would, once it had arisen, immediately recede back into the unconscious, in other words it could not even exist for itself' (Merleau-Ponty, 1962, p. 206).

Naming something does not come after the recognition of the thing but constitutes its recognition: the word itself is the founder and bearer of meaning, rather than a mere vehicle. Through bodily expressions such as words and gestures we are nearby the things in the world. However, the meaning of a gesture does not occupy the gesture as a physiological quality. Likewise, the meaning of a word does not inhabit the word like the sound of the word does. Expressions transcend their physical existence by extending beyond themselves towards the world. As such, words and gestures create an intersubjective horizon of meaning within which we share a common world. Similar to Heidegger's notion of the poetic word, drawn out of silence, Merleau-Ponty calls for attention to the creative power of language:

> *As soon as man uses language to establish a living relation with himself or with his fellows, language is no longer an instrument, no longer a means; it is a manifestation, a revelation of intimate being and of the psychic link which unites us to the world and our fellow men (...). It might be said, restating a celebrated distinction, that languages or constituted systems of vocabulary and syntax, empirically existing 'means of expression, are both the repository and the residue of acts of speech, in which unformulated significance not only finds the means of being conveyed outwardly, but moreover acquires existence for itself, and is genuinely created as significance. Or again, one might draw a distinction between a speaking word and a spoken word. The former is the one in which the significant intention is at the stage of coming into being (...). Speech is the surplus of our existence over natural being. But the act of expression constitutes a linguistic world and a cultural world, and allows that to fall back into being which was striving to outstrip it. Hence the spoken word, which enjoys available significances as one might enjoy a required fortune. From these gains other acts of authentic expression – the writer's, the artist's or philosopher's – are made possible. This ever-recreated opening in the plenitude of being is what conditions the child's first use of speech and the language of the writer, as it does the construction of the word and that of concepts* (Merleau-Ponty, 1962, pp. 196–197).

Speaking of subjects and objects ends here, because this speaking veils the primordial reality in which they are united. Only in their intentional engagement with the world, can human beings every now and then raise their heads above the things around them. Only in such temporary situations, can subjects and objects be said to stand in opposition to each other. However, even then they are both rooted in the same reality: the flesh of the world. Therefore, this 'intentional dualisation' should be seen as emanating from an original belonging and togetherness: *'j'en suis'*.

3.4 CONCLUSION

Many parallels can be drawn between Noddings' phenomenology of care and Merleau-Ponty's phenomenology of perception. Apart from the most evident parallels – the primacy of the life–world, the primordial status of human relatedness, the questioning of the autonomous subject – I would like to point to the fact that both phenomenologists reveal a fundamental reversibility between acting and undergoing; analogous to Merleau-Ponty's reversibility of perceiving and being perceived. Noddings stresses our caring and being cared for as two sides of the same, reversible practice. This reversibility manifests itself in the fact that it is in our experience as *cared-fors* that we develop the fundamental caring intuitions that enable us to be responsive *ones-caring*. This reversibility is displayed as well in the fact that mature and 'sustainable' relations are marked by mutual caring, in which the parties exchange places; both members are *carers* and *cared-fors* as opportunities arise. However, whereas Noddings conceives of an exclusive kind of reciprocity, restricted to human relations, Merleau-Ponty reveals a carrying reciprocity that is operative in our involvement with the perceived life–world in general. While the insights of Merleau-Ponty are worthy of discussion in their own right, I want to confine myself here to the contribution of his phenomenology of perception to our understanding of ecological responsibility. More specifically, does Merleau-Ponty's understanding of the reciprocal relationship between human beings and their world allow for an extension of the caring imperative to the natural environment? First and foremost, the phenomenology of perception allows us to criticise some of the ontological assumptions that underlie Noddings' denial of such an extension.

In the first place, Merleau-Ponty's understanding of intercorporeality provides us with strong arguments to question the sharp dichotomy that Noddings draws between the existential necessity of our engagement with other humans and the contingency of our engagement with the natural environment. This dichotomy is false, since it insufficiently accounts for our sensual and bodily situatedness, our physical immersion in the fleshly world. From an ontological point of view, our engagement with the surrounding natural life–world is equally given with our being in the world as our engagement with other humans. As body–subjects, we partake of both worlds. Merleau-Ponty uses the term intercorporeality to indicate the primordial interconnectivity between me and my natural world that precedes individual consciousness and can therefore not be disconnected by an individual agent. I exist only insofar as I am committed to this world and involved in its affairs. Just as we cannot cut our 'self' off from the significant others that carry our self, so too we cannot cut off our 'self' from the natural world in which we are intentionally involved and of which we partake as fleshly beings. This acknowledgement of an ontologically necessary involvement in our life–world clears the way towards an extension of the ideal of caring towards the natural world. After all, our social situatedness as well as our natural situatedness sustains our being in the world in an equally fundamental way.

In the second place, and intimately tied to the above, Merleau-Ponty's understanding of the dynamic reversibility – the mutual reflection and implication of perceiving and being perceived – sheds new light on the reciprocal relation we maintain with our natural world. This reciprocity is beyond the scope of Noddings' ethics of care, since she chooses to take the interhuman caring relation as paradigm model for all our caring. From the perspective of human caring, Noddings observes a shading-off from the ethical into the sensitive and aesthetic, as she discusses our relationships to plants, things and ideas. Here we are left with personal sentiments, attachments and aesthetic pleasures that are free of obligations. Thus Noddings acknowledges that one's engrossment in these 'things' might engender an *I-must*, but not in an ethical sense, since our care is not received and answered in a way that sustains my ideal self as one-caring. I agree that, indeed, it makes sense to stress that the reciprocity we experience in our bodily involvement with the natural life–world will be of a different kind than the reciprocity we experience among human beings. And these differences are partly of such a fundamental nature that they have implications for the possibility of ethical caring for nature. Things, plants and animals may respond to my caring gaze and involvement, but this responsiveness is limited to a certain level of complexity. As Noddings rightly remarks in relation to the caring relationship with her cat: she 'is a responsive cared-for, but clearly her responsiveness is restricted: she responds directly to my affection with a sort of feline affection – purring, rubbing, nibbling. But she has no projects to pursue. There is no intellectual or spiritual growth for me to nurture, and our relation is itself stable. It does not posses the dynamic potential that characterises my relation with infants' (Noddings, 1984, p. 156).

Our care for nature is therefore obviously of a different kind than our care for persons. However, Hans Jonas argues (along with Kant) that, while non-human organisms do not pursue ideals, they are purposive, in the sense that they are driven by life-preserving instincts: organisms strive towards their own survival and flourishing. Our care for nature is partly premised by the sense that in order to flourish, our flourishing depends on the flourishing of other organisms in our life-world in a non-instrumental way. In other words, the flourishing of our natural world is constitutive for our flourishing, just like the flourishing of our beloved ones is. As body–subjects we partake, as it were, in communal forms of flourishing. I care for the well-being of my intimate friends, not merely for instrumental reasons, but for intrinsic reasons: because their well-being and our relationship are in itself valuable to me, are part of me, in a way I cannot deny, even if I wanted to. According to Jonas, the same goes for our care for nature (Jonas, 1984; O'Neill, 2001).

The empirical fact that interhuman reciprocity is not operative in our relation with the natural environment does not imply that there is no other form of reciprocity between us and our natural life–world that causes an ethical imperative of caring to exist, to which we must respond by virtue of the very fact that we are human. The precise nature of this reciprocity and its ethical implications are in my view revealed by Merleau-Ponty in his examination of the chiasm between humans and their world: the reversibility of perceiving and being perceived. Our intentional involvement in the natural world brings with it a particular kind of receptivity for the appeal that emerges out of this involvement: the worldly things that I perceive and experience, reveal to me who I am. Thus, I change and the things that appear in my world change, as soon as a common field of vision unites us. In this sense, there is an authentic, reflecting reciprocity between me and my natural world, analogous to the reflection taking place in interhuman relations. Therefore, unlike Noddings suggests, we should understand our involvement with the natural world as one of the sources of our self. The question of who we are cannot be answered without reference to where we are and how we relate to our natural life–world. Consequently, if I take responsibility for my self – for the person I am – I simultaneously assume responsibility for the sources of my self, among which my practical involvement with nature. In other words: care for the self implies care for my natural life–world.

The parallels between the phenomenology of care and the phenomenology of perception bring to light a striking similarity between the experiences of caring and perceiving, that allows us to suggest that caring and perceiving are rooted in a common mode of being in the world. Moreover, I would like to suggest that we understand the reciprocal relation established by intense perception as a caring relation. The phenomenological kinship and common ontology of caring and perceiving justify the question as to whether we should conceive of our corporal and sensual engagement with the natural life–world, and our openness towards this world, at the same time as a caring involvement with the natural life–world. We find support for this suggestion within the phenomenological tradition of thinking, starting with Heidegger's notion of care as the human mode of being (*Sorge*) and carried on by philosophers like Hannah Arendt, Hans Jonas and Merleau-Ponty. What these phenomenological thinkers have in common, is that they do not see the ability of rational autonomy as the distinguishing hallmark of human beings, but our initial openness to the world around us. Since we are always already in the world and engaged with its affairs, we do not have to establish our contact with this world by artificial or technical means. Rather, the things that appear in our world already speak to us through their immediate presence. For us, the rake is not merely an instrument, a technical artefact, but an extension of my body, my arm. The rake is absorbed in our daily habits and practices, even before I take an objectifying distance towards my actions and ask myself what its function is and which use would be most effective. The meaning of this rake is expressed in my acquaintanceship and familiar use. As such, its meaning is extracted from the particular life–world of which the weeder, the spade, the watering can and young lettuce in my kitchen garden are all a part as well. This world embodies a personal web of meaning that is not deliberately spun by me. Moreover, the things in this web are not accessible to us by command but display their full meaning only insofar as we are passively involved in this life–world and are receptive to its presence. We are commissioned to be together with the worldly things in a way that enables these things to appear in their authentic being. This attitude of solicitousness toward other things and this concern to do things meticulously embody what Heidegger calls a caring involvement. The receptive being in the world is a caring being in the world (cf. Taylor, 1995; Noddings, 1998; Jonas, 1984).

Such a caring mode of being does not primarily imply responsibility for pristine or virgin nature, untouched by humans – located at the largest possible distance of our cultural world. Rather, it is our 'common-or-garden' care for nature in our immediate vicinity, being absorbed in our everyday practices and routines, that generally makes up this caring attitude. Our care for nature shows itself in the engagement we involuntarily express within the pre-reflexive rituals and movements of our bodies. Not so much the spectacular climbing of a mountain top or the once-in-a-lifetime dive in the barrier reef, but rather the weekly weeding in my backyard, the daily cooking ritual or the cycling trip to and from work might give rise to the expression of such a caring imperative in our behaviour. It is in my familiar contact with the kitchenware, the ingredients, litter, water, natural gas, electricity and kitchen waste, that the internal *I-must-care* arises. The same goes for cooking. For instance, many people will experience a certain care for the ingredients in their tactile involvement with the ingredients – the washing of lettuce, the peeling of potatoes, the kneading of dough or the shelling of beans. Thus, it is said of a good cook that he treats his ingredients with respect so that the ingredients come to fulfil their promise. But even apart from this promise of taste, my practical involvement with the ingredients fulfil me with a feeling of pleasure (cf. Smith, 1998c).

The caring involvement with the things in our life–world is in a sense 'commissioned' by these things, rather than determined by our preferences. It is to this kind of 'attunement' that Richard Smith refers in his characterisation of the practical reason of craftsmanship as a powerful antidote to the instrumental means-end calculation, pervasive in mechanised, repetitive and alienated labour:

> *The hallmarks of practical reason are flexibility and attentiveness to the details of the particular case (Aristotle calls this attentiveness aesthesis, sometimes translated as 'perception' or 'situational appreciation'). It is coloured by sensitivity and, crucially, attunement towards its subject-material, rather than the attempt to exercise mastery or control over it. The craftsman, for instance, has a certain 'feel' for the wood or stone he is working, and knows that if forced it will split or shatter. Instrumental or technical reason produces goods which are specified by criteria that lie outside the process of making. The car that comes off the assembly-line is determined by considerations of what can be sold to the customer; the manufacturer is unlikely to be moved by the thought that a different way of going about the process will help to keep alive certain craft-skills among the workforce. In practical reason, on the other hand, we seek the good that we attempt to realise through the action and not as a separate and independently identifiable aim (…). In this way practical reason is irreducibly ethical, and its ethical quality is of a rich and complex kind, involving a continuous, if not always fully conscious, testing of one's*

> *action against the internal goods of an activity (...). The carpenter knows his wood and has respect for its qualities. The experienced cook employs her knives, pans and other equipment so to speak as an extension of herself and comes to know quite instinctively which flavours complement which. Like all craftsmen the carpenter and the cook must be patient, methodical, sometimes extemporising ingeniously. This of course is to talk of those in a position to practise their craft with a sufficient degree of autonomy and creativity. It is precisely because the fulfilling work in which these crafts can be practised has so widely been replaced by mechanised, repetitive and alienated labour (or 'drudgery') that we both forget how readily available these possibilities of 'attunement' have been until recently and at the same time incoherently romanticise the craft traditions in which they are found* (Smith, 1998c, pp. 174–175).

In this light, it is important to recall that we do not care for nature because we want nature to remain as it presently is but because we want to preserve our worldly practices. Any understanding and appeal that emerges from our practical involvement with nature is, in a sense, cultural. For some nostalgic minds this might cause a feeling of regret and loss, but as Hannah Arendt shows, the etymology of the word 'culture' suggests that it is this cultural bearing that allows us to take care of our natural world:

> *While other species rely on instinct to develop nature into their habitat, human beings have to depend on culture. The traditional work of culture, indeed, has been to make nature into our habitat or home (...). The word 'culture' derives from colere – to cultivate, to dwell, to take care, to tend and preserve – and it relates primarily to the intercourse of man with nature in the sense of cultivating and tending nature until it becomes fit for human habitation. As such it indicates an attitude of loving care and stands in sharp contrast to all efforts to subject nature to the domination of man* (Arendt, 1961, pp. 211–212).

Conclusively, the main argument in favour of an alternative understanding of ecological responsibility in terms of care (as distinguished from the mainstream understanding of responsibility in terms of choice) proceeds as follows: just as we know and value ourselves and our actions in dialogue with others, so do we gain self-understanding and self-respect from our pre-reflexive practical involvement with the natural environment: the landscape, the skies, beloved animals and plants or the 'green grass of home'. The particular form of reciprocity underlying this involvement – a reversibility of perceiving and being perceived – allows us to speak of a caring relation. This means that our care for nature is received and answered by nature as a *cared-for* in a way that sustains our self-understanding as *ones-caring*: the blooming flowers in spring answer my continuous care for my garden. Or even to a lesser degree: the black soil blinking back at me after my weekly weeding session pleases me in a way that makes my caring involvement fulfilling. We care for nature immediately at hand

and relate to the things around us on a pre-reflexive level. Analogous to Noddings' derivation of an ethical ideal of caring from this 'natural caring', the remembrance of ourselves as *ones-caring* is of essential importance in situations in which our care for nature is not intuitively felt but nevertheless required by our ideal. We should call upon our ideal self as caring person and thus activate our latent desire to care for our natural environment. It is along these lines of an evolving ethical self, to which we are committed in one way or another, that an *I-must* emerges.

This *I-must* is no longer a pure contingent and intuitive one, but an ethical imperative to which we must respond since it is premised by our very self-understanding and feeling of human dignity. A critic might respond that we can perfectly develop a sense of one's self and a sense of what it means to be human without any involvement in our natural life–world whatsoever. I doubt this, because as humans, we are all biologically predisposed to eat, drink, sleep and find shelter against the cold; we are in a sense conditioned to make ourselves at home in our natural life–world. This is to say that our life sustaining functions condition us to interact with the natural environment in order to eat, drink, sleep and so on. It is due to the caring nature of our human being in the world, that we cannot relate to these functions in a purely functional manner: once we eat and drink our involvement with nature can not be restricted to mere metabolism – a mutual exchange of chemicals – but this interaction brings with it an acquaintanceship with the natural environment. Trust is, for instance, an important element in our consumption of food. We would never eat – let alone enjoy eating – if we do not trust the food that is (made) available. Accordingly, we would never catch sleep if we were unable to trust our natural environment in a minimal sense. Thus, a certain acquaintanceship and relation of trust with our natural environment is given along with the biological necessity to eat, drink, sleep and find shelter. We simply have to inhabit our natural world in order to sustain our human form of life.

The strength of our ideal of ecological responsibility as originating from an evolving ethical self, has to show what it is capable of when confronting world-wide problems of global warming, economic inequality, bio-diversity and so on. After all, our caring intuitions are insufficient when it comes to meeting the responsibilities we assume as world citizens, since these global problems exceed the limits of our life–worlds. In short, the problem is how to understand the connection between our caring intuitions that are intimately tied to our immediate life–world and the responsibilities we assume as global citizens. Indeed, it is not hard to show that our local actions will have impact on a global level and *vice versa*, that our local action is shaped by global conditions, structures and agreements. There are causal chains of action consequences in directions, exceeding far beyond the present and the particular here and now. With this empirical in mind, my question is more specific: how can we draw on local intuitions and commitments in our global adherence to environmental responsibility? I would like to respond to this question along two lines that are nevertheless intimately related to each other.

In the first place, my horizon of caring is extended by means of the previously outlined appeal to my ideal self as a caring person. On the strength of my past caring experiences – the care for the robins and titmouses in my backyard – I know myself as a caring person, or more specifically, as someone who cares for nature. And it is because I am attached to this self-understanding that I am able to call upon this image of myself as a caring person and act in accordance with the sense of duty – the ethical *I-must* – that arises from this understanding. Ergo, it is my ideal self that commands me to care for something or someone in contexts where my natural caring responses fall short. Thus, we sometimes experience the influence of a moral imperative that involuntarily forces itself upon us, in situations in which we would not be 'naturally' be inclined to care. Of course, this duty will be internalised in such a way that it comes to be natural caring intuition as well. Thus, our moral and political sensitivities extend beyond the limits of my life–world. How far our ideal of caring takes us, depends primarily on the ability and strength of our imagination. When considering the purchase of tropical hardwood, for instance, are we able and prepared to imagine the beauty and natural richness of the Amazon rain forests? Are we able and prepared to place ourselves in the position of the aboriginal inhabitants of these forests and imagine the injustice that is done to their community life by western importers, local large landowners and corrupt politicians?

My point is not that caring persons necessarily privilege their environmental commitments above others, because indeed, we might experience counter-commitments that make an equally strong demand on us: our esthetical imagination and appreciation of a hardwood window-sill in my living room might fulfil me with such a strong feeling of comfort that it outweighs any ethical ideal of caring. Rather, my point is that it is indeed possible to draw on our locally situated ideal of caring in global matters by extending our ethical caring intuitions in such a way that we can imagine how the consequences of our actions will affect natural entities, human practices and communities elsewhere and in future. Whether such an ideal of caring responsibility will be strong enough to deal with the global problems with which we are confronted will largely depend on the way in which this imagination is cultivated in our common practices, so that its members will inevitably gain a particular sensitivity to environmental considerations of a global nature. Throughout our participation in caring practices we might develop a moral sensibility that urges us to care for foreign and future nature *as if* it were ours. As environmentally caring citizens we act in line with the ideal that this hypothetical image elicits within us. I will come back to the practical implications of this analysis in the last section, in which the familiar environmental adage *Think Global, Act Local* will be discussed in more detail.

In the second place, it might be fruitful to think of this extension of our caring commitments in terms of expanding circles and chains, as Noddings suggests:

> *We find ourselves at the centre of concentric circles of caring. In the inner, intimate circle, we care because we love (…). As we move outward in the circles, we encounter those for whom we have personal regard. Here, as in the more intimate circles, we are guided in what we do by at least three considerations: how we feel, what the other expects of us, and what the situational relationship requires of us. Persons in these circles*

do not, in the usual course of events, require from us what our families naturally demand, and the situations in which we find ourselves have, usually, their own rules of conduct (…). Beyond the circles of proximate others are those I have not yet encountered. Some of these are linked to the inner circle by personal or formal relations. Out there is a young man who will be my daughter's husband; I am prepared to acknowledge the transitivity of my love. He enters my life with potential love. Out there, also, are future students; they are linked formally to those I already care for and they, too, enter my life potentially cared-for. Chains of caring are established, some linking unknown individuals to those already anchored in the inner circles and some forming whole new circles of potential caring. I am "prepared to care" through recognition of these chains (Noddings, 1984, pp. 46–47).

We may think of the expansion of our caring sentiments toward organic and animal life, future generations and natural beauty as proceeding along the lines of these circles and chains, the recognition of which will cause one to be prepared to care. Thus, particular imaginations grappling with how the consequences of our (in) actions will affect other humans, animals, landscapes and natural systems might transform this passive preparedness into an active caring for the natural environment. Contrary to the chains of rights-approach, suggested by Rawls and his interpreters, this latent concern for distant and future flourishing does not originate from a predefined contractual relationship, assuming a strict reciprocity between moral parties, but an enlarged acquaintanceship with widening circles of life. Thus understood, the limits of our responsibility cannot settled in advance, but they reach out in directions where our reflecting sensitivities to life deepen. As such, this understanding of environmental responsibility comes close to Naess' notion of ecological selfhood and self-realisation. Rather than a self that employs the environment as a resource for his own purposes, Naess postulates a self that matures by means of identification with ever wider circles of being. For Naess, nature begins to have intrinsic value by virtue of the fact that self-realising persons enlarge their 'selves' in such a way as to encompass nature and become one with it (Naess, 1973; Passmore, 1974; Noddings, 1984, p. 152; chains of love; Wenz, 1988; Hargrove, 1992; Li, 1996; Matthews, 2001; Wenz developed a similar concentric circle theory of environmental justice).

To sum up, we derive an ethical ideal of caring responsibility for the natural environment along the following lines:

(1) We experience our practical–corporeal involvement with the natural world as commissioning us to respond as *ones-caring*: our sensual and bodily being in the world is of such a nature that, within our pre-reflexive involvement with the natural things as they appear in our life–world, we involuntarily display a certain care for these things.

(2) It is in (anticipated) dialogue with significant others that we recognise the natural things we care for and we reflect on our caring identifications with the natural world in terms of the strong-evaluations that we borrow from our *sensus communis*.

(3) Throughout the continuous involvement with nature and throughout our continuous dialogue with others we gain a particular self-understanding and, thus, an ethical self evolves in congruence with our best remembrance of nature calling on us to care: I feel myself committed to the idealised understanding of myself as a person that cares for nature.
(4) When we experience an appeal to our caring responsibility – by virtue of our recognition of chains of care – even though our intuitive caring responses are nevertheless absent – since the call is made from beyond the limits of our present life–world – we call upon our ideal self as caring persons in order to see how far our ethical ideal of caring will take us. We extend our caring sensibilities toward organic-, animal-, distant- and future-life along these concentric lines, but obviously, this extension of our horizon of caring goes as far as it goes.
(5) The strength of our ethical ideal of caring for the natural environment depends largely on the degree of institutionalisation and cultivation of our care within the practices of our political community as *sensus communis*. Only on the strength of these practices can the extension of our caring sentiments be sustained and internalised by subsequent newcomers.
(6) Our ethical ideal of caring responsibility for nature is always susceptible to change, since it responds to the call that emanates from our continuous involvement with nature. After all, the value of nature transcends our ethical ideal of caring in the sense that nature resists any human appropriation within a fixed and self-fulfilling understanding of ourselves as caring persons.

NOTES

[1] This is my translation of the poem BWA-PL of Willem Jan Otten, from his collection *Eindaugustuswind* (End-of-august-wind): *Wij bereikten/ na een tocht door een druipend bos/ het Randmeer./ Het was alsof een slapende haar ogen opende/ en ons kende/ Jij zat voorop./ Ik legde mijn hand/ Op de warme kokosnoot van je schedel/ Het licht keek ver in je ogen/ Ik zei: dit is nu water./ Wa-ter/ Wa-ter/ Wa-ter, zei ik nog een keer/ En jij zei: bwa-pl/ Je zei het nog een keer/ Het was zeker, zoontje van mij, dat wij hetzelfde niet begrepen* (Otten, 1998).

[2] In pre-modern times, this existential awareness was signified as a profound feeling of piety, about which Roger Scruton writes: 'Put in simple terms, piety means the deep down recognition of our frailty and dependence, the acknowledgement that the burden we inherit cannot be sustained unaided, the disposition to give thanks for our existence and reverence to the world on which we depend and the sense of the unfathomable mystery which surrounds our coming to be and our passing away. All these feelings come together in our humility before the works of nature and this humility is the fertile soil in which the seeds of morality are planted (…). Piety is rational in the sense that we all have reason to feel it. Nevertheless, piety is not, in any clear sense, amenable to reason. Indeed, it marks out another place where reasoning comes to an end. The same is true, it seems to me, of many moral attitudes and feelings: while it is supremely rational to possess them, they are not themselves amenable to reason, and the attempt to make them so produces the kind of ludicrous caricature of morality that we witness in utilitarianism' (Scruton, 1998, p. 50).

[3] Translation of a phrase in: Seel, M. (2000). *Ästhetik des Erscheinens*. Munchen. This translation is borrowed from the website of the German Goethe Institute: http://www.goethe.de/kug/prj/kan/en81442.htm.

[4] Idem.

[5] Inspired by Levinas and Llewelyn, Paul Standish articulates this insight in terms of otherness: 'the relation to the Other is not realised in a kind of abstract contemplation but rather in language itself' (Standish, 2003, p. 109).

[6] Consider for instance the phrase in which Rorty expresses his reluctance towards the word 'truth' because its use mostly indicates 'a way of allowing a description of reality to be imposed on us, rather than taking responsibility for choice among competing ideas and words, theories and vocabularies' (cited from: Bonnett, 2003, p. 597).

[7] Consider for instance the twelfth principle of the eighteen principles of sustainability as summarised in the *Rio Declaration on Environment and Development* (1992), and rehearsed in the *ESD Toolkit* (2002): 'Nations should cooperate to promote an open international economic system that will lead to economic growth and sustainable development in all countries. Environmental policies should not be used as an unjustifiable means of restricting international trade' (McKeown, 2002, p. 9).

[8] This quote is borrowed from the *Philosophy of Education Yearbook* (1995), published on-line at http://www.ed.uiuc.edu/eps/pes-yearbook/95_docs/marshall.html.

[9] Idem.

[10] Idem.

[11] Cuypers speaks of concepts of autonomy in stead of concepts of responsibility.

[12] In fact, Cooper elaborates on the Wittgensteinian insight – expressed in the previous chapter – that reason comes after the belief and that a change of belief is not a matter of rational persuasion alone. Moreover, a change of belief requires openness to 'see things differently' (as commissioned by the dawning of an aspect).

[13] Merleau-Ponty primarily aims at a particular kind of empiricism, called sensualism. The term 'rationalism' is taken to refer to the dogmatic rationalism à la Descartes, though Merleau-Ponty criticises the rationalist elements in Kant's transcendental project as well.

[14] In fact, 'être-au-monde', as Merleau-Ponty writes, should be translated as 'being-towards-the-world' but within ordinary language as well as phenomenological English 'being-in-the-world' is more commonly used.

CHAPTER FOUR

BECAUSE WE EDUCATE CITIZENS CARING FOR NATURE

'Give me your definition of a horse' (...)

'Girl number twenty unable to define a horse!' said Mr. Gradgrind for the general behoof of all the little pitchers. 'Girl number twenty possessed of no facts in reference to one of the commonest of animals! Some boy's definition of a horse. Bitzer, yours' (...)

'Bitzer', said Mr. Gradgrind. 'Your definition of a horse'.

'Quadruped. Graminivorous. Forty teeth, namely twenty-four grinders, four eye-teeth, and twelve incisive. Sheds coat in the spring; in marshy countries, sheds hoofs, too. Hoofs hard, but requiring to be shod with iron. Age known by marks in the mouth'. Thus (and much more) Bitzer.

'Now, girl number twenty' said Mr. Gradgrind, 'You know what a horse is"

Charles Dickens[1]

Children are thrown into a world whose existence is at stake. Though the human world has always been at stake, now its survival is being threatened by the creations of its inhabitants. Human beings turned out to be vulnerable to the consequences of their collective behaviour. By creating an aggressive-technological world, by installing an exploitative economic system and enjoying consumer society, we are unwillingly endangering the condition of future life on earth. Furthermore, the threats to our world are now of such a magnitude and nature that the danger is not solely limited to a particular area, but puts the very 'health' of our global environment at risk. Problems of environmental degradation, pollution and resource depletion far exceed the local horizons of community action. Haunted by alarming future scenarios, local as well as global efforts are undertaken to 'sustain' the things we care for by changing our practices, behaviours and creations in such a way, that they will be less threatening to the continuation of our worldly practices.

As children are introduced to our worldly practices, they will become familiarised with these efforts to take up responsibility for the quality of our natural life–world and habitat; as they learn how to deal with waste and energy within their family household, at school, at their first job, in leisure time, on the street and in public debate. As newborn citizens, they are involved in collective practices as fellow-citizens, who gradually come to participate and simultaneously assume co-responsibility for the continuity of our practices in the future. But what precisely do those educators do when they involve children in our practices and simultaneously strive for continuity of those civil practices? Are environmental educators offering an appealing perspective on a possible future world, or do they treat their pupils as human material for the establishment of this future world to come? In others words, does the claim of continuity do justice to the right of children to an open future horizon, or does it lead to a colonisation of our future in such a way that we deny new generations the opportunity of shaping their own future world? Philosophers of education generally warn against the utilisation of education as a means of creating some preconceived 'ideal society'. On the other hand, every educational act and judgement casts their shadows ahead. Every educational intention or response lays claim to a particular future of indeterminate length. By acting in a particular way now, the educator commits himself and the child to doing something in the future. Moreover, our action as well as inaction whether we or not we act has consequences for the world we leave behind for future generations. Therefore, we cannot remain innocent and a total refraining from continuity claims in education can not be logically maintained.

In chapter two we concluded with Hannah Arendt that this need for a guided introduction calls upon the double responsibility of environmental educators; by virtue of their position as mediators between an old world and new generations, educators need to be responsive to the implicit claims of young generations to an open future, as well as to the implicit claim of our 'old world' in need of protection. In fact, educators maintain a relation based on trust with the young as well as the old; they require the young to trust the old and require the old to trust the young by protecting the newness of the young against the established powers of the old and protecting the established world against the destructive newness of the young. In Arendt's view adults are only able to fulfil their twofold task by introducing the young into our world as it is at present is, in all its potential and with all its flaws. Only when educators act as representatives of our present world, for which they assume responsibility, will the fundamental conflict between both claims be dissolved. This is because in order to preserve our shared world it has to be continuously renewed. And from the perspective of the young reform is only possible when newcomers take their bearings from the present world in which they are properly introduced. So, rather than being utopian prophets environmental educators should be guardians of our common world.

If future responsibility emerges from our commitment to what we care for here and now, rather than our hypothetical relationship with future citizens, how do we stand towards the things we value? Do we 'make' our natural environment valuable by ascribing value, or do we 'find' value in nature? In chapter three, an alternative view has been developed on 'the intrinsic value of nature' in terms of aesthetic judgement. In this view, the value is neither intrinsic to the subject, nor to the object of valuation, but resides in the common language that enables us to speak about nature, based on experiences to which we are all subject. As such, the value of nature is embedded in the intersubjective level of our language community. However, as I have argued, the intrinsic value of nature transcends our articulation of value. Its ultimate value and meaning eludes us.

As we have seen, the issue of intrinsic value is important to our understanding of what it is that inspires us to care for nature i.e. the caring responsibility we assume in our practical involvement with the natural environment. Nel Noddings' phenomenology of care as well as Maurice Merleau-Ponty's phenomenology of perception has allowed me to describe the emergence of an ethical ideal of caring responsibility for the natural environment. In my view, this ideal should be preserved within the educational practices and arrangements of our society. But of course, any appeal to environmental responsibility will have to connect to the playful involvement with nature, in which children maintain an intrinsic engagement with their natural environment. In this concluding chapter I will therefore start out by exploring the moral and educational significance of this playful involvement with the natural environment. Apart from this dimension of play, I will draw attention to two more dimensions of which I will argue that they are of vital importance for environmental education in the light of the previous analyses: self-expression and the primacy of the natural life–world. In line with this primary characterisation of environmental education, I will call for a revision of the dominant framework of *education for sustainable development* (ESD). Finally, an alternative proposal will be presented, to be called *education for environmental responsibility*.

Child's play

Taking a glance at the list of dispositions required for an aesthetic recognition of intrinsic natural value, one can hardly resist the impression that we are dealing with serious, adult matter here. Weighty terms like 'dwelling', 'contemplation', 'poetic involvement', 'receptive responsiveness', 'acquaintanceship' and 'attunement' allude to adult experience rather than children's involvement with nature. After all, children seem to approach their natural environment in a typical childlike way. They experience nature mainly as something to play with and play in. Apparently, nature appeals to their spontaneous responses of imaginative creativity. Thus, trees challenge children to climb in them, ditches invite them to build dams, sand to make castles out of them, the moon to lure werewolves, woods trigger children to play hide and seek, rabbits invite them to care for them and chestnuts to gather them and make funny creatures out of them. But nature does not merely appeal to responses of a noble kind. At the same token, animals sometimes release a kind of sadistic playfulness within children. Consider for instance how cruel children can be towards vulnerable animals – putting salt on snails, pulling out the paws of a spider, chasing cats – eager as they are to explore what it feels like to be superior and exercise power. Though we might question the moral quality of this behaviour, in these sadistic responses children display an intrinsic involvement as well: the animals are tormented without any purpose, but 'just for fun'.

There is no consensus among researchers in the field of education and developmental psychology on what play is mainly for and about. Moreover, research literature on child's play shows that it is impossible to give one definition of child's play that covers the various dimensions of this phenomenon. Rather, various (partly overlapping) characteristics are given. The more these characteristics apply to particular behaviour, the more likely an observer will regard this behaviour as play. First, the characteristic of intrinsic motivation refers to the fact that an action is done for its own sake and not brought about by basic bodily needs or by external rules or social demands. Second, the activity of playing is pleasurable and enjoyable to the child. Third, the behaviour is non-literal, meaning that it has an 'as if' or pretend quality. Fourth, the flexibility stands for the amount of variation in form or context that playing usually displays (these characteristics are borrowed from Smith and Vollstedt, 1985). Apart from these main characteristics, there is attention for the dimensions of voluntariness, bodily immediacy, the degree of engrossment in the play, joy in repetition, and the enchantment of reality (cf. Gadamer, 1960, pp. 97–116; Langeveld, 1963; Huizinga, 1971; Imelman, 1974; Hutchison, 1998, pp. 104–105).

All listed characteristics of play reflect particular dimensions of the aesthetic experience of intrinsic value, as outlined in the previous chapter. First, the characteristic of intrinsic motivation reflects the dispositions of the aesthetic appreciator or the one-caring, as previously characterised in terms of motivational displacement, disinterestedness and engrossment. Second, the characteristic of joy can be seen as the childlike version of the personal pleasure and fulfilment that appreciators of nature find in aesthetic gratification. Thirdly, the characteristic of 'non-literalness' correlates with the creative power of imagination that is at the heart of aesthetic judgement. Fourth, the flexibility of play exemplifies the continuous adjustment of responsive evaluators to the time- and place-particularity of the aesthetic experience. Finally, the bodily immediacy in child's play refers to the sensual intimacy between aesthetic appreciators and their object of care.

These parallels indicate that within free play, children as well as adults, realise an aesthetic involvement with the environment: a tacit responsiveness to the things that appeal to them. Interrelationships between play, aesthetics and self-expression become even more evident as we discuss the ideas of Hans-Georg Gadamer, who defines play in relation to the experience of a work of art. However, for Gadamer, play is not so much a subjective exercise of imagination but something (a movement of being) that has its own order and structure to which the one playing is handed over. This movement of play is not under voluntary control of the one playing. Therefore, Gadamer speaks of the primacy of the play over and above the player. Every act of play is an act of being-played: 'Alles Spielen ist ein Gespielt-werden' (Gadamer, 1960, pp. 108–110). Paradoxically, it is within this responsive play that human beings display themselves, or more precise: within this play they are pushed to express themselves in a particular way. At this point, Gadamer points at the analogy between self-display, human play and nature:

> *Das Spiel stellt offenbar eine Ordnung dar, in der sich das Hin und Her der Spielbewegung von selbst ergibt. Zum Spiel gehört, daß die Bewegung nicht nur ohne Zweck und Absicht, sondern auch ohne Anstrengung ist. Es geht wie von selbst. Die Leichtigkeit des Spiels, die natürlich kein wirkliches Fehlen von Anstrengung zu sein braucht, sondern phänomenologisch allein das Fehlen der Angestrengtheit meint, wird subjektiv als Entlastung erfahren. Das Ordnungsgefüge des Spieles läßt den Spieler gleichsam in sich aufgehen und nimmt ihn damit die*

Aufgabe der Initiative ab, die die eigentliche Anstrengung des Daseins ausmacht. Das zeigt sich auch in dem spontanen Drang zur Wiederholung, der im Spielenden aufkommt und an dem beständigen Sich-Erneuern des Spieles, das seine Form prägt (z.B. der Refrain). Daß die Seinsweise des Spieles derart der Bewegungsform der Natur nahesteht, erlaubt aber eine wichtige methodische Folgerung. Es ist offenbar nicht so, daß auch Tiere spielen und daß man im übertragenen Sinne sogar vom Wasser und vom Licht sagen kann, daß es spielt. Vielmehr können wir umgekehrt vom Menschen sagen, daß auch er spielt. Auch sein Spielen ist ein Naturvorgang. Auch der Sinn seines Spielen ist, gerade weil er und soweit er Natur ist, ein reines Sichselbstdarstellen (Gadamer, 1960, pp. 110–111).

Careful phenomenological study of child play gives rise to the idea that children express within their play the fulfilment they find in an aesthetic involvement with nature that allows them to recognise (in a practical and immanent way) the intrinsic value of nature. Of particular interest within the context of environmental education is the magical 'animism' inherent in the child's play world. In particular, young children between the ages of four and twelve respond to nature in an empathically and anthropomorphic way, ascribing human-like emotions and desires to animals, plants and things. Nature triggers their imagination in a special way; organisms and things are involved in their play as if they were active and intentional beings. Pets are taken care of as if they were babies in need of parenting; trees moving in the dark before dawn are regarded as ghosts; sick flowers are being nurtured and spoken to as if they were patients; the holes in the sand on the beach are underground cities that have to be protected against the marching breakers that come rolling in. According to Margadant-van Arcken, who studied child play in nature, such a response is not an expression of animistic reasoning (suffering from the 'pathetic fallacy' or a 'category mistake'). Children who display animistic responses in their play, do not necessarily assume that flowers have human-like feelings and that trees are bewitched. Moreover, their imagination allows them to express the appeal arising from their intimate involvement with nature in animistic terms (Margadant-van Arcken, 1990, p. 149). This is a clear manifestation of Merleau-Ponty's reversibility between perceiver and perceived; young children are not yet aware of emotions, thoughts and intentions as 'inner experiences' as things that can be judged independently of the world at which they are directed. An interesting result of research into the cognitions of young children is that a sense of desire that belongs to us more than to the world at which they are directed, only emerges with the progressive acquisition of knowledge (Margadant-van Arcken, 1998).

Some environmental philosophers (implicitly or explicitly) suggest that anthropo*morphism* can be seen as inversely proportional to anthropo*centrism*. (Lemaire, 2002; Bonnett, 2003). The imagination that allows us to recognise human-like intentional forces beyond ourselves literally pushes our 'selves' away from the centre of our world. Thus the assumption is the more one responds to these forces in an empathic way, the less narcissistic our desires, views and attitudes will be. Support for the thesis that anthropomorphic animism enhances our care for nature can be found in manifestations of animism in mediaeval culture, its origin myths and the underlying metaphor of nature as Mother. Like a real mother, Mother Nature was believed to suffer from changing moods and tempers to which her children had to adjust. Thus, belief in Mother Nature required respect and dictated a subservient attitude of piety. By satisfying Mother Nature and worshipping natural gods, mediaeval people believed they were in good hands; protected against natural disasters and supported by good fortune. In tribal cultures as well as in the contemporary Gaia belief, a similar animism is cultivated as source of wonder and respect (cf. Bonnett, 2003, pp. 580–581).

Obviously, my suggestion is not to retrieve a version of pre-modern mysticism within our present concept of nature, nor to transmit Gaia doctrines within practices of environmental education. Rather, my aim is to reveal the ethical meaning of child play in nature. Though more detailed qualitative educational research is necessary to sustain my thesis, the previous exploration provides us with strong reasons to assume that the kind of aesthetic involvement in nature that allows us to respect the intrinsic value of nature, can be found in child play in a most spontaneous and imaginative form. Within child play educators recognise that they do not have to teach children to care for nature by means of transmission. They only need to appeal to a caring-bodily responsiveness that is present in any child.

However, having acknowledged the importance of play, it is worrying to see that the opportunities for free play in nature are decreasing in contemporary western education. Educational research shows that children spend less time in nature (Biologieraad, 2002). However, these studies tend to define nature rather narrowly in terms of natural reserves, while nature in the broad sense we have discussed inevitably resides in the child's life–world. Indeed, there is a major difference between the quality of nature in the life–world of a farmer's child and the environment of a child growing up in the suburbs of Amsterdam. But even in big cities, environmental educators have succeeded in marking out exciting nature footpaths and creating green play areas as well as school gardens. In light of this, I think that the main educational problem is not the availability of 'play nature' in the environment of the child, but rather the opportunity to play in nature. For, obviously, the relationship between children and their life–world has been transformed into an ever more controlled and mediated relationship. Unstructured time and opportunity to play, more or less free from adult supervision, is rarer now, than it was ten or twenty years ago. It is not my purpose here to give an extensive outline of the causes and roots of this loss of free time in contemporary education, but undoubtedly, the loss of free time and opportunity to play is closely related to changing priorities in contemporary education and upbringing. For instance, the growing concern of parents for their child's safety, the desire for monitoring children's development and the felt importance of efficient use of educational time have diminished these opportunities. To give a concrete example: the playgrounds of contemporary child care centres in the Netherlands are subjected to severe safety standards that will not allow any 'risks' in the child's environment. As a consequence, trees are protected from being climbed, branches are removed since children might hurt one another, differences of height are surrounded by safety measures to prevent children from falling, and even the smallest ditch is relegated from the child's surroundings. Thus, the illusion of a risk-free playground deprives children of many opportunities to be triggered by raw, natural material in which to explore their bodily powers and simultaneously explore the powers of nature. Paradoxically, because of these safety measures, children are becoming even more vulnerable, since they have not learned to deal with risks anymore. They are not able to judge whether or not it is safe to jump down from such and such a height; they are not used to estimate how far they can move into the water before the current will catch them, and so on (De Valck, 2004, pp. 11–13).

While schools cannot compensate for society, I do think that they have a special educational assignment here. By making room for an alternative-playful involvement with nature, schools are able to counterweight the performative tendencies within society. However, I agree with those critics of environmental education who stress that educators should resist external claims to contribute to a predetermined approach to social problems. In virtue of their own task, purpose and position in society, educational institutions enjoy a certain degree of autonomy against political and economic institutions in society. Educational institutions would sacrifice their primary responsibility and expertise were they to give in to thoughtless expectations from the outside (Meijer, 1996, 1997; Praamsma, 1997; Bolscho, 1998; Margandant, 1998). Among other things, educators should not simply adopt the current, scientific and technological definitions of environmental problems, presented to them by environmentalists, policy-makers and scientists, because these definitions sometimes fail to connect to the life–world experience and environmental awareness of their pupils. Moreover, the special responsibility of educators amounts to the translation of social problems in such a way, that they connect to the purpose and meaning of the educational practice. Ergo, when it comes down to enhancing environmental responsibility, it is not a mere social or economic task that schools are burdened with but an educational assignment: educators ought to preserve the playful and caring involvement of children with their natural environment and inspire them to deal with the caring responsibilities that arise from this experience.

Unfortunately, the particular forms of environmental education enhancing intimate physical interaction with nature – such as nature education, nature study and conservation education – have been pushed to the back within the framework of ESD. In fact, many contemporary organisations for environmental education want to get rid of the image of nature education, as they feel this is practice is intimately tied to the outmoded image of the 'open sandals and woolly socks' type of education. Against protagonists of ESD, I have stressed the aesthetic and ethical importance of child play. In line with this analysis, I want to call for a revaluation of nature education, or – more neutrally – education in nature. This dimension of environmental education should provide children with the learning environment and opportunity to play in and with nature, free from immediate educational interference, evaluation and curricular purposes.

Self-expression

By setting limits to the satisfaction of our consumer needs, environmentalists have, willingly or unwillingly, contributed to the promotion of a morality of austerity and self-restraint, as if they want to say: all things in life that are pleasant, comfortable, tasty or exciting are bad for the environment, and should therefore be subject to rigorous self-discipline. In particular, advocates of sustainable development tend to underline that, if we want to live a sustainable life, we have to deny ourselves the pleasures of consumption and control our unsustainable impulses and bad habits. After all, the ideal of a sustainable development sets limits to our present use of natural resources, in such a way that these limits are assumed to be attributable to individual consumer behaviour, as defined in our 'ecological footprint'. Consequently, we should be willing to make our 'shallow desires' – our desire for a long morning shower, for cheap meat and vegetables, or for frequent holidays by air – subordinate to our desire for a sustainable future. Assuming environmental responsibility is presented as the ability of self-discipline or self-restraint that allows us to resist the temptations of consumer culture.

In itself, this picture of environmental responsibility is recognisable and seems to be realistic, but, as I have suggested earlier, it reflects all the shortcomings inherent in the ideal of rational autonomy and the underlying notion of self-identification in terms of choice and accountability as outlined in the previous chapter. Furthermore, a strong appeal to self-restraint and self-discipline fails to touch upon our intrinsic motivation. Instead it primarily trades on our controlling responses, fuelled by fear; fear for loss of the things we care for, fear for future catastrophes, fear for our survival or the survival of posterity, fear for falling short in the eyes of others, fear of losing control of our future action and so on. As previously discussed, this insight lead Jonas to acknowledge the 'heuristics of fear'; more precise and stronger than positive mental states like joy or desire, fear of specific future dangers points our attention to those vulnerable goods we really care about (Jonas, 1984, p. 27). It might sound far fetched to mobilise fear in such a way, but in fact, many environmentalists have gained support and credibility by sketching dramatic future scenarios: 'If we do not change our ways drastically within a short period of time … .'. However, in the long term such a strategy has proven to be a dead end. The fact that in hindsight some of the darkest doom scenarios have appeared to be exaggerated has given some environmentalists the dubious image of being modernday Cassandra's, that is, prophets of doom to which nobody listens anymore (but as Fukuyama once remarked, the position of the pessimist is, in a particular sense, a privileged one; when the prophecies of optimists turn out to be too bright they easily obtain the stigma of being gullible and naive, whereas pessimists retain the aura of profoundness when prophecies have proven to be false).

In my view, Jonas is right to stress that fear can be an indicator of what we care about and for, but I agree with those educational theorists who argue that fear should not be used as an educational means to motivate children to act, think or judge in a particular way. That would imply a step back to the dark ages of puritan education, in which the awareness of hereditary debt and the fear of a punitive God were the pedagogical Leitmotif. In a similar vein, Noddings argues against those existentialist thinkers who elevate fear – fear of death, fear of meaninglessness or fear of human existence as such (Heidegger or Sartre) – to the status of our true mode of being. Against the heuristics of fear, Noddings and Margadant-van Arcken, for instance, suggest that educators ought to connect to childlike play and care and joy in life in their moral appeal. Rather than assuming responsibility for fighting the things we fear, we should take responsibility for preserving the things we care for and about. This is not just a matter of definition, but a difference in quality and will be clear as we take a look at the implied engagement of the subject.

According to Kant's *Critique of Judgement* (1790), it is through our sensuous pleasure in beauty that we are able to transcend the inner conflict between virtue and desire: '(...) in the enactment of aesthetic perception we are in a special way free – free from the compulsions of conceptual recognition, free from the calculation of instrumental action, and free from the conflict between duty and inclination' (Seel, 2004; cf. Taylor, 1992, p. 71). Accordingly, the previous analysis of intrinsic natural value in chapter three provides us with profound reasons to assume that the aesthetic experience of natural beauty allows us to move beyond our desire for self-discipline towards a personal striving for aesthetic self-creation, understood in terms of self-expression: the self as a work of art, as Foucault chooses to say. Whereas the hierarchical notion of responsibility in terms of rational autonomy and choice starts with the necessity of making one's first-order desires subordinate to our desires of a more profound, higher order – thus appealing to our pursuit for self-control – the alternative notion of responsibility in terms of care allows us to be engrossed by the things for which we care; to lose ourselves in an object of care. This caring is of an aesthetic nature, in the sense that it is an experience of responding to something particular that appeals to us. In our caring we express the response to something particular that fulfils us in a particular way, and invites to follow: 'When the other's reality becomes a possibility for me, I care' (Noddings, 1984, p. 14). As such, this appeal to environmental responsibility takes on the form of an appeal to responsive self-expression; an invitation to express what we desire in our involvement with the natural world, to express the kind of life we want to lead in harmony with an image of ourselves and our world which we feel comfortable with.

Inherent to these two different practices of environmental responsibility are different styles of self-evaluation. In qualifying this difference, it might be helpful to employ Charles Taylor's distinction between weak calculative self-evaluation and strong qualitative self-evaluation. Suppose that a particular person wonders how she might change her means of transport in order save money and 'save' the environment. Now, this person might weigh the alternative options by calculating which behaviour changes will contribute most effectively to her objective of saving, and simultaneously require the least sacrifices on her part. Biking to the workplace in stead of driving will probably yield a reduction of greenhouse emissions. Simultaneously it might enhance my 'health' and other personal goods, but unfortunately, it will conflict with the pursuit of 'comfort' and 'saving time'. After weighing up the relative importance of these personal standards – ranking 'sustainability', 'saving money', 'health', 'comfort' and 'saving time' in the order from most important to least important – she might be able to reach a decision. She might choose to opt for a minimal change in her ways – only cutting out one of her holidays by air – since more radical changes would 'cut' too much into her feeling of 'comfort', 'saving time' and other goods. This type of evaluation fits Taylor's notion of 'weak calculative self-evaluation': weighing alternatives on grounds of pre-specifiable standards in quantitative terms: option × satisfies my standards more/less than option y. The goal is to determine the option which produces the greatest amount of over-all satisfaction.

According to Taylor this calculative picture of self-evaluation fails to come to grips with the richness of human judgement and the moral horizon within which we are situated. As such, we do not reflect on our actions primarily in quantitative terms but in strong qualifications that are interwoven with our self-understanding and way of life. In this stronger style of self-evaluation the actor qualifies her desires not merely in quantitative measures on the scale of 'strongest/weakest' or 'most/least important', but as being worthy or unworthy, virtuous or vicious, more or less fulfilling, more or less refined, profound or superficial, noble or base. These strong evaluations are part of a vocabulary of qualitative contrast: the qualitative terms gain meaning by contrast with other qualifications. Thus, it is impossible to explain what we mean when we call someone or something 'profound' without reference to qualifications like 'shallow' or 'superficial'. Consequently, changing the use of one of these terms, or introducing a new term would alter the sense of the existing terms, thus colouring the entire vocabulary. Suppose, for instance, that someone starts to use the word 'profound' in an ironic way, then the connotations of 'shallow' and 'superficial' will change apace. Furthermore, strong evaluations of this kind belong to qualitatively different modes of life; 'profound' is not just an abstract qualification but alludes to a way of life to which we might or might not aspire, or it alludes to the kind of person we might or might not aspire to be.

Evaluating her means of transport in this stronger sense, the person in our previous example would reflect on the available options in qualifications that amount to the way of life she as a person feels right. She might, for example, recognise the option of biking to her work as 'inspiring' and 'deepening contact with nature' by reminding herself how 'refreshing' it is to experience the gradual change of the landscape, to become familiar with the trees and plants along the road, to experience the cycle of seasons. Rather than balancing costs and benefits according to general standards that can be settled in advance, the personal vocabulary of worth she employs is invoked by considerations of this stronger qualitative kind. These qualifications are different from the general standards of the 'weak evaluator' in at least one major aspect: the standards of the weak evaluator can be defined independently of the particular considerations at hand. After all, saving money or realising a reduction on energy use and greenhouse emission can be realised independent of a change in her travelling habits. These goals may be realised by other means as well. In case of strong qualitative evaluation the qualifications I use are not independent but intrinsically related to the particular consideration at hand: the person in our example may feel that her life in harmony with nature might eventually be disrupted if she does not leave her car for a bike. Furthermore, it follows from this that, when weak-calculative evaluation is practiced,one desired alternative is set aside, it is only on grounds of its contingent incompatibility with a more desired alternative or another goal. But with strong qualitative reflection this not the case; the conflict is deeper. The aspiration to live in harmony with my natural environment does not compete with the person's desire for saving time, since she values the former aspiration as more profound than the latter. Moreover, the higher aspiration to live in harmony with nature might change her aspiration to save time in a qualitative manner; she may, for instance,come to reject her desire for saving time as a desire belonging to a superficial, hurried life she does not want to live anymore. As such, strong self-evaluation seems to leave less room for self-deception since it is the very self that is at stake (Taylor, 1969, p. 283).

In strong qualitative self-evaluation, the person evaluates herself and simultaneously articulates what she feels is important or valuable in life. It is clear that self-evaluation in this strong sense is not only articulated but constituted by a particular kind of language and expression. Rather than being mere descriptions, strong evaluations are articulations of the kind that do not leave the object of evaluation unchanged: 'To give a certain articulation is to shape our sense of what we desire or what we hold important in a certain way'. Again we touch upon the aesthetic style of self-expression enabling us to transcend the conflict between external principles and internal desires by giving in to the urge of self-creation. Charles Taylor observes how such a qualitative development of the self emerges within our radical re-evaluations:

> *Because articulations partly shape their objects (...), they are intrinsically open to challenge in a way that descriptions are not. Evaluation is such that there is always room for re-evaluation. But our evaluations are the more open to challenge precisely in virtue of the very character of depth which we see in the self. For it is precisely the deepest evaluations which are least clear, least articulated, most easily subject to illusion and distortion (...). The question can always be posed: ought I to re-evaluate my most basic evaluations? Have I really understood what is essential to my identity? Have I truly determined what I sense to be the highest mode of life? (...) But in radical re-evaluations the most basic terms, those in which other evaluations are carried on, are precisely what is in question. It is just because all formulations are potentially under suspicion of distorting their objects that we have to see them all as revisable, that we are forced back, as it were, to the inarticulate limit from which they originate. How then can such re-evaluations be carried on? There is certainly no meta-language available in which I can assess rival self-interpretations. If there were, this would not be radical re-evaluation. On the contrary the re-evaluation is carried on in the formulae available, but with a stance of attention, as it were, to what these formulae are meant to articulate and with a readiness to receive any gestalt shift in our view of the situation, any quite innovative set of categories in which to see our*

> *predicament, that might come our way in aspiration. Anyone who has struggled with a philosophical problem knows what this kind of enquiry is like. In philosophy typically we start off with a question, which we know to be badly formed at the outset. We hope that in struggling with it, we shall find that its terms are transformed, so that in the end we will answer a question which we couldn't properly conceive at the beginning (...). The same contrast exists in our evaluations. We can attempt a radical re-evaluation, in which case we may hope that our terms will be transformed in the course of it'* (Taylor, 1969, pp. 296–298).

By positioning consumer needs in opposition to our responsibility for the use of natural resources the principle of sustainability gives far too much weight to our needs as consumers. This is not to say that we should deny, neglect or suppress these needs, but to suggest that we could understand them differently by revaluating those needs as part of our desires of a more profound nature; our desire to live sincerely, our desire to lead meaningful lives, or our desire to realise ourselves as caring persons that find fulfilment in their involvement with nature. As these consumer needs are reframed within the horizon of existential desires they are likely to change in quality: the need of 'saving time' may transform into a desire to find rest in one's life and one's activities. As such, a strong qualitative evaluation changes the object of evaluation in such a way that our responsibility no longer consists of 'sticking to our principles' against immediate inclinations, but in the creation of a personal style of involvement with nature that renders beauty and meaning to our life. Thus, the conflict between principle and inclination dissolves in favour of a constitutive desire to live differently, to be a different person, or to live in a different world. The recent emergence of a *Slow Food Movement* illustrates the social nature of these desires; people get together around shared commitments to those practices of farming, craftsmanship and cooking within which our intrinsic care for nature is preserved.

If we take seriously this understanding of self-evaluation within the practice of environmental education, we may conclude that it is not appropriate to instruct children in the meta-language of sustainability and challenge them to change their behaviour according to its meta-principles; putting one's inclinations under the guidance of an abstract ideal of intergenerational responsibility. These principles are powerless and meaningless unless they are embodied and 'modelled' by dedicated persons – educators – and integrated in everyday practices. Rather, environmental educators should challenge children to explore their present involvement with nature, to dwell on the corporal dimension of this involvement and express their experiences in conversation with others. Personal stories and autobiographical essays could play a part in this expression. Subsequently, pupils and students question one another's behaviour and self-expression in terms of the motives and desires that are inherent to them. This questioning should not be like a quest for justification or mere accounting for the effects of one's behaviour, but like an existential appeal to each other's (well) being-in-the-world: How do you stand towards the things you care for, the things you despise, the things you are highly indignant about, the things you strive after?

How do you respond to these things and who are you in relation to these things? The practical forms of such a conversation in schools will be discussed in my alternative proposal for *Education for Environmental Responsibility*.

Primacy of the natural life–world

In previous analyses on the nature of environmental responsibility we have repeatedly felt the fundamental tension between the *local* experience of nature and the assumption of responsibility for *global* problems of climate change, biodiversity, pollution and degradation. Whereas the sources of environmental care and responsibility are located (mainly) within the social life–world, the objects of our responsibility are generally identified from a global perspective. Thus, the subject is torn between two worlds: care for the natural life–world that is immediately at hand and responsibility for the global environment that we can only picture by means of abstract scientific language. One way of dealing with this tension is expressed in the environmental phrase 'think global, act local', often referred to by advocates of sustainable development. Though it is not exactly clear what 'global thinking' is taken to mean – as it is generally employed without further explanation in an apparently self-evident way – it is possible to distillate some of its meaning from the agreements of *Agenda 21* (1992), in the context of which this slogan is most often used.

First of all, the relationship between the local and global dimension of responsibility is framed in terms of accountability for the transfer of harmful consequences: the consequences of our actions for the environment and well being of people elsewhere and in the future are no longer to be 'externalised' from the calculations that inform local choices about the use of natural resources. Second, in *Agenda 21* great emphasis is placed on the 'translation' of the global language of sustainability to local practices and communities. Local governments, institutions, associations and individuals are expected to participate in the promotion and implementation of sustainable development in a way that meets local needs and interests. Thus, the exact implications of sustainable development are not settled in advance, but determined in an ongoing conversation and process of adaptation to changing circumstances. Third, on the level of knowledge, the primacy of global thinking amounts to an over-all favouring of scientific knowledge. Likewise the relation between local and global knowledge is understood in terms of deduction and induction. It is tacitly assumed that the kind of knowledge that will enable us to solve environmental problems is causal-analytic knowledge about the causes and effects of pollution, depletion and degradation. By monitoring the global transfers of these environmental effects in terms of costs and benefits, we will be able to extract general laws that can be deduced to local practices of decision making, so that their policies are 'evidence based', thus advocates of sustainable development assume.

Judged against the background of globalisation processes and the ever further reaching consequences of human behaviour in a technological age, it is a matter of course that the environmental agenda is partly determined by global concerns. Environmental educators cannot afford to resort to mere local concerns in order to create a green *Ecotopia*, without considering the condition of globalisation. Within the practice of environmental education, children have to be sensitised to the ways in which their involvement with the natural life–world is being regulated by social, economic, technological and political structures and institutions. However, I do object to the uni-linear and deductive conceptualisation of the relationship between local and global concerns within the leading language of sustainable development. If 'global thinking' implies that local concerns are to be framed in global terms of sustainable development, if 'global thinking' amounts to a narrow ideal of being accountable for the global consequences of one's actions against the background of fixed standards, or if 'global thinking' is about positioning oneself as a footprint-holder in a global system of interdependencies, then local concerns are conceptualised as a mere function of global concerns. My analysis of the relationship between particularity and generality in political judgement (from the perspective of Arendt) in chapter two gives rise to a reverse picture. Local care for our natural life–world precedes global responsibility. According to this aesthetic understanding of political judgement, particular concerns for our life–world cannot be subsumed under a general rule, but the concerns themselves reveal a general idea that we will never be able to articulate exhaustively.

In this context, Arendt points at the *exemplary* validity and power of judgements within political discourse. Judgements that appeal to particular events sometimes reveal a strong general idea that cannot be expressed otherwise. Thus, particular events in recent history have fuelled public debate on environmental issues and shaped our awareness of environmental problems in an unprecedented way. Accordingly, the intrinsic values that are at stake in a particular local concern of environmental care will be repudiated if they are immediately subsumed under a global rationality of accountability. Intrinsic qualities are preserved only if our thinking proceeds from the particular to the general. Thus, ideally, 'anticipated communication' starts from within those local discourses in which the particular significance immanent to this concern appears. As local concerns raise questions of a more general nature, we move on by anticipating wider circles of (imagined) audience as well, however, without ignoring our initial audience. Eventually, this 'enlarged way of thinking' may touch upon global issues of sustainability and development, but this is not necessarily so. Some concerns can be dealt with on a community scale or within national arenas of political discourse. Thus, in a sense, we should always start off by 'thinking locally', taking multiple perspectives of all participants involved in a particular concern about the natural environment in order to highlight what is immanent about this particular concern. If these local problems turn out to be of such a nature, that their significance exceeds the limits of the local discourse community (i.e. that the problems are of a strong systemic nature), then obviously, action and reflection on a more general scale is necessary. Thus, the rate of generality depends on the exemplary validity of a particular concern.

Here, I will outline my criticism of the 'primacy of the global' more precisely along the lines of two particular objections: the first objection concerns the privileged status granted to scientific knowledge in the curriculum, and the second objection amounts to the inherent pursuit of a consumerist freedom of choice. Simultaneously I will suggest an alternative approach to these issues.

In the first place, global discourses on environmental problems as well as school curricula are mainly informed by scientific knowledge of nature, thus privileging causal explanatory knowledge above other forms of knowledge, such as life–world knowledge we possess through bodily contact with the world; a way of knowing that 'senses the immanent in the particular' (Bonnett, 2003, pp. 649–650). Obviously, we cannot do without science and technology in the approach of environmental problems, since most of these problems only become visible by means of scientific methods or instruments. But the fact that environmental problems can be identified by science does not imply that these problems can be fully comprehended by means of science. Rather, there are different ways of knowing that are equally valuable. Some of the alternative ways of knowing may even be regarded as more profound, such as the pre-reflexive bodily understanding of our natural life–world Merleau-Ponty speaks of. This understanding of nature is primordial against the scientific abstractions that are authoritative within environmental discourse. Scientific knowledge is necessarily derivative of this primordial acquaintanceship, hence Merleau-Ponty. Because of this derivative nature, environmental educators should be vigilant with scientific concepts. It is for instance important to note that the 'globe' as such is highly counterintuitive to us, as Spivak argues: 'You walk from one end of the earth to the other and it remains flat. It is a scientific abstraction inaccessible to experience. No one lives in the *global* village (…). My question, therefore: in what interest, to regulate what sort of relationships, is the globe evoked?' (Cited from Bonnett, 2003). For many, the global perspective seems to hold promises of manipulability: by taking an Archimedean standpoint from outside of the globe, it is like we are in the position to regulate the earth's natural processes and determine everyone's place and responsibility accordingly. Obviously, these promises of global engineering are illusive and express a misplaced feeling of arrogance and control.

Recognition of the derivative nature of scientific knowledge gives cause to a revaluation of the status of scientific knowledge within the school curriculum in general, and practices of environmental education in particular. If we understand knowledge about nature as a mere function of the active knowing subject, then we lose the intuitive sense of nature as an independent part of our life–world and a source of imagination and beauty (cf. Arendt, 1958b, pp. 257–267). The monopolisation of our understanding of the environmental crisis by science should be counterbalanced by meaningful life–world experiences in order to elicit a response of care and commitment within children. Unfortunately, translation from life–world knowledge to scientific knowledge and *vice versa* appear to be a perilous undertaking. According to Margadant-van Arcken, these problems are due to the fact that children tend to have a biotic view of nature – composed of green plants, trees and animals – as opposed to an abiotic view of the environment, supported by a grey picture of plumes of smoke in the air and toxic waste in river water. These two views are separated from one another, so that most children hardly recognise the interconnections between their practical concerns of nature and abstract concerns of the environment. In order to establish a transfer between life–world knowledge and scientific knowledge that goes both ways, Margadant-van Arcken suggests that it is important to acknowledge the primacy of the life–world. Her research shows that, when children are confronted with new phenomena, they resort to 'life–world thinking' before they move on to more abstract scientific language. So, only if they get the opportunity to familiarise themselves with the new by dwelling, comparing and reflecting on these new phenomena in the midst of familiar life–world phenomena, are children more likely to internalise scientific understanding of natural processes and environmental problems. As such, the natural life–world operates as the very substratum of scientific knowledge (Margadant-van Arcken, 1998).

A similar suggestion is made by Bonnett, who equally underlines the 'priority of the local': '(…) for the global to achieve significance it needs to be affectively as well as cognitively rooted in the local. This means that what is at issue is no longer simply a set of international generalisations produced by disengaged rationality, but the carrying forward of an acquaintanceship with the local into the global' (Bonnett, 2003, p. 655). By familiarising ourselves with 'the global' scientific knowledge becomes part of our life–world. This familiarisation does not proceed along the lines of deduction from general principle to particular knowledge. Rather, it emerges in a different kind of generalisation from the particular 'here' and 'now' to 'there' and 'then', drawing on the strength of imagination. Thus, scientific concepts like 'life cycles', 'acid rain', 'greenhouse effect' and 'food pyramid' are expressive in such a way that they have rapidly become part of our everyday conversations and expressions. Other terms like 'photosynthesis' and 'eutrophication' have been less successful in pervading ordinary language, but perhaps they can be relabelled in more expressive terms, so that they speak to us non-scientists in a more appealing sense.

In the second place, I would like to counter the 'primacy of the global' by questioning the inherent adherence to a consumerist freedom of choice. As previously indicated, by taking an Archimedean standpoint from outside of the globe, we do not only harbour the illusion that we are in the position to regulate the earth's natural processes. Moreover, this position also leads us to think that it is possible to determine everyone's place and responsibility accordingly; from our global knowledge about the limits of the earth's resources we can deductively calculate the maximum use of natural resources per individual, i.e. our ecological footprint. Unfortunately, global processes of atmospheric relations and climate change are far too complex and unpredictable to be monitored in this way. Having acknowledged this, many scientists and environmentalists nevertheless seem to anticipate this ideal of transfer from a global definition of the environmental crisis to the settlement of individual margins of consumption. However, as long as evidence is not sufficiently firm, these settlements cannot be imposed onto us without conflicting with our basic liberty rights. Furthermore, it is important to underline that global discourses are marked by scientific language in tandem with the global language of free market capitalism. This means that efforts to organise environmental responsibility on a global level are distributed along the lines of liberty rights and consumer freedom. These rights are envisaged in terms of freedom of choice. Thus, global institutions like the Worldbank and International Monetary Fund (IMF) require national governments to privatise their energy markets, so that people can choose whether they opt for sustainable energy or regular energy generated by fossil fuels. The main idea behind these liberalisations (often implying or resulting in privatisations) is that green products have to prove themselves by generating their own public.

The rhetoric of sustainable choice promises individuals that they will be able to realise their personal freedom by making informed choices in accordance with their own preferences and self-chosen standards of sustainability. However, in a free market environment such freedom of choice comes to operate as a disguised vehicle of uncritical self-restraint, since – as indicated earlier – we are only able to choose insofar as we subject ourselves to the regime of a consumerist choice industry. In such a global system our 'free choices' are regulated in directions that optimise its performativity. Thus, we are made co-responsible for 'choices' that were enforced by the system. Here, the 'self-as-a-reflexive-chooser' has become subject to ideological manipulation. This is what I would like to call the hidden assumption of manipulative citizenship, underpinning many discourses of sustainable development. If politicians, policy-makers and captains of industry fail to internalise the environmental costs in the prices of products and services, they tend to call for an education that teaches citizens to choose for the option that is officially labelled as the most 'sustainable' one. In this light, the appeal to 'think global, act local' may be understood as appealing to an uncritical application of globally defined principles on a local level. 'All in all you're just another brick in the wall' (Pink Floyd). As such, ESD easily comes to serve as a tool to shirk responsibility in the hands of those political authorities and institutions who fail to take environmental responsibility themselves. Furthermore, our critical horizon is being narrowed by distracting attention from the political issue to the private issue of personal choice.

An alternative counter-practice to this top–down approach to environmental education – operating as a vehicle for the promotion of sustainable development – is most likely to emerge from the world-wide exchange among pupils, students and scholars on the internet, who work on environmental issues in their own region or country. Rather than being some kind of abstract ideal, international educational projects like *Globe* (Global Learning and Observations to Benefit the Environment)[2] and *Codename Future*[3] are supporting pupils to investigate the quality of soil, air, water and vegetation in their school environment, and share their findings with scientists and children all over the world (cf. Wijffels, 2004, pp. 18–19). By sharing their 'local' knowledge and experiences, they will be able to explore the global conditions of environmental responsibility. What are the institutional possibilities and constraints for environmental action? Through joint reflection on local practices, possibilities for collective action may open up. Thus, new forms of solidarity have emerged around the anti-globalisation movement, based on shifting coalitions of environmentally spirited (future) citizens who met on the web and joined powers in particular political struggles. Different from collectives based on a universally shared ontological condition – the working class, women and gays – these collectives originate from a 'groundless solidarity', thus Edwards and Usher argue. Participants are brought together on the basis of a shared commitment concerning a particular issue, but without explicit claims to inclusiveness. With respect to another political issue, participants might just as well proceed in divergent ways again, thereby making room for new coalitions. Thus, coalitions and meeting places grow more contingent and global processes of political participation will become less predictable. As soon as one starts acting in global affairs one will be immersed in the capricious dynamics of global politics. As a consequence of this 'immersion' in a complex and interminable web of meanings and power relations, the agent loses control over the meaning of his words and actions. The consequences of our actions proceed in directions we can not possibly predict. Calculating what the effects of local action will be on global decision making will become ever more difficult. Therefore, once more, imagination will have to do the job (Arendt, 1958b, pp. 236–247; cf. Edwards and Usher, 2000, p. 134).

However, as outlined in dialogue with Arendt, environmental education should not be designed to educate environmental activists: 'Exactly for the sake of what is new and revolutionary in every child, education must be conservative' (Arendt, 1958b, pp. 510). Moreover, educators are to familiarise children with the problems that threaten our shared world as part of their introduction into that world, thus Arendt argues. Only when educators refuse to take responsibility for these worldly problems, and take refuge in utopias, do risks of indoctrination and manipulation lie in wait. Therefore, it is important to familiarise future citizens with the global politics of sustainable development and the neo-liberal free market-policies. It is in this world that children will have to develop their own stance towards the things in life they value. Their ability to make judgements is only likely to develop within a world of contrasts, oppositions and imbalances of power. This argument for introducing children into our present world with all its potential and all its flaws should not be understood as a call of commitment to neutrality. On the contrary, in the way educators act and teach they inescapably express what they care for in this world and the ideals by which they live (cf. Carr, 2004, p. 221). But their caring should always be presented towards children as a question mark, as a possible mode of being, as an appeal to make up their own minds, rather than an educational effort to transmit our caring commitments and ideals to them. In many cases this may be the result, but transmission of ideals should never be the intention of those educators who sincerely claim to educate to foster independence of judgement. In this spirit, Heyting argues that 'offering ideals to children does not make them ideals for children (…). In order to become an ideal for a child, she needs to develop a personal, existential, commitment to it; she needs to (re)create the ideal as such. Making sure she feels free and stimulated to do so, seems the overriding aim with respect to the role of aims in education. Too much attention to the substantive ideals that educators prefer – however thoroughly tested – may even hinder children from creating their own, if only for fear of disappointing their educators' (Heyting, 2004, p. 246).

It is important to reiterate my point that the proclaimed primacy of the life–world does not imply that we can simply ignore the globalising processes and structural conditions of environmental responsibility. On the contrary, we should reflect thoroughly on these global dimensions and extend the limits of children's life–world in ever wider circles towards the global world, so that they become familiarised with worthwhile practices on the other side of the globe and imagine what 'goods' are at stake in the environmental problems that we identify. To be more precise, there are three particular ways in which the tensions between local and global concerns are to be dealt with in environmental education. First, children are persuaded to consider the consequences of our present (in)actions towards global processes of climate change, depletion, pollution and degradation and future practices here and elsewhere, not primarily by means of calculation, but through imagination. Second, children are stimulated to explore the ways in which their actions are regulated – constrained and made possible – by global institutional structures, forces, policies and free markets. In this context Giddens speaks of 'utopian realism': examining the prevailing institutional practices for immanent but neglected and as (yet) unused possibilities to give new impulses to our caring responsibility for the natural life–world (Giddens, 1990; Jansen, 1994, p. 258). Third, by making use of the possibilities offered by internet (e-mail groups, e-learning, e-meeting, e-publishing, e-government) children are inspired to share their local experiences and results of environmental exploration and action with children, scientists and politicians elsewhere. In this triple sense, children move from the local life–world to the global environment and backwards in ways that do not privilege particular forms of knowledge or ways of being in the world.

ESD revisited

Though it is not my purpose to refute the language of sustainable development as such – because I think it has proved its worth by bringing together disparate stakeholders and providing them with powerful incentives to cooperate in contexts where institutional interests conflict – I do think it is necessary to move beyond the current exclusive framing of any appeal to our caring responsibility for the natural environmental within the language of sustainable development. Previous elaborations on the dimensions of child's play, self-expression and the primacy of the natural life–world, allow me to specify my objections against the framework of ESD in more practical–educational terms. First, the dominant framework of ESD tends to underestimate the fundamental importance of a playful interaction with nature. Thus, it fails to make room for a personal sensitisation to the appeals arising from our intimate involvement with the natural environment. After all, it is through emergent acquaintanceship with their natural life–world that children develop a caring commitment to protect, sustain and preserve nature.

Second, within the framework of ESD, the process of taking responsibility is generally understood as a matter of appealing to personal accountability, i.e. holding one accountable for the consequences of one's actions, measured in terms of pre-specified margins of consumption. By positioning consumer needs in opposition to our responsibility for the use of natural resources, the principle of sustainability gives far too much weight to our needs as consumers. As such, proponents of ESD have, willingly or unwillingly, contributed to the promotion of a morality of austerity and self-restraint, as if they want to say: all things in life that are pleasant, comfortable, tasty or exciting are bad for the environment, and should therefore be subject to rigorous self-discipline. Such a strong appeal to self-restraint and self-discipline fails to touch upon our intrinsic motivation, but instead primarily trades on our controlling responses, fuelled by fear. In our aesthetic experience of nature, on the contrary, we are able to transcend the inner conflict between virtue and desire, by giving in to the urge of self-expression.

Third, the framework of ESD fails to connect global concepts of responsibility to our local caring and acquaintanceship with nature. By advocating the priority of 'global thinking', ESD seems to privilege or favour those forms of knowledge, that are informed by scientific and economic discourses on environmental problems above other forms of knowledge, such as poetic knowledge and life–world knowledge we possess through bodily contact with the world around us. By narrowing environmental education in the curriculum to a hardcore of predominantly causal-analytic knowledge, ESD largely ignores the primordial nature of the life–world as our existential source of knowledge, values, attitudes and caring commitments. Recognition of the derivative nature of scientific knowledge gives cause to a revaluation of the status of scientific knowledge within the school curriculum in general, and practices of environmental education in particular.

While I feel sympathetic towards those critics who have broadened the meaning of the leading concepts of sustainable development, sustainability and development, I do not share their (self-) evaluation; I do not believe that it is within the power of an individual author to intervene in our common use of language in such a way it gets all the participants in a particular practice to understand sustainable development in a radically different way. Those introducing a new meaning of sustainability and arguing that their version of sustainability has nothing to do with sustainable development, overestimate their power. Authors who oppose the common meaning of 'sustainability' or sustainable development cannot simply choose to use the word in their own proposed new meaning without any consideration for the common use of language within a particular practice or language community on which the reception of their terms depend. It is obvious that they can, but their words will be not be understood in a way that reflects the author's intentions. Their words will be framed in terms of the dominant discourse. For instance, I have outlined that the central concept of 'development' can hardly escape connotations derived from the free market economy. Couched in such terms, economic development is immediately interpreted as economic growth, which in its turn is equated to the maximisation of profits and capital accumulation. For this reason I do not subscribe to the proposals of those who suggest that we should understand sustainable development in a completely different way than is commonly defined.

Education for environmental responsibility: a proposal

My criticism of the underpinnings of the current framework of ESD gave rise to an alternative understanding of environmental responsibility and education, as previously expressed in terms of play, expression and the primacy of the natural life–world. However, suggesting an alternative understanding of an educational practice is one thing, proposing a change in practice is quite another and requires a different kind of knowledge. My research is not primarily concerned with the empirical practice of environmental education. Consequently, I lack the empirical knowledge about the educational activities that are actually taking place under the labels of environmental education and ESD in everyday school life; this knowledge is prerequisite for any authoritative suggestion of practical change in environmental education. However, I have closely studied the dominant frameworks(s) of sustainable development in which politicians, policy-makers, environmentalists and educators evaluate the activities in the field of environmental education and design new ones. My philosophical criticism on these frameworks has informed my proposal for an alternative framework. As with any conceptual framework, this proposal does not directly amount to practical change but to a change of understanding and evaluation. While such a change is likely to bring about practical changes, this is not necessarily so.

It is against the background of my educational and philosophical criticism on the dominant language of sustainable development that I want to present an alternative framework for environmental education: *education for environmental responsibility*. The nature and outline of this proposed framework is generally in line with the conclusions of my philosophical inquiry but, as with any practical proposal, it is not determined by these conclusions, and surely not the only possible realisation of its suggestions *in concreto*. More exemplifications would be possible. Nevertheless I feel that philosophers of education should not only aim for rigorous philosophical analysis, but should, eventually, stick out their necks and express, as engaged educationalists, what they stand for[4].

Within the framework of *education for environmental responsibility* environmental education is understood to be an introduction into those collective practices in which our involvement with the natural environment is somehow captured and expressed in a way that inspires children to recognise the things that sustain our bodily-perceptive sense of aesthetic fulfilment as caring beings in the world. To be more precise, this introduction will sensitise children in a playful way to the experience of an appeal to care that originates from our practical involvement with the natural environment. Moreover, this introduction will familiarise children with stories, histories, games and vocabularies that challenge them to express their care for the natural life–world in a personal way. Furthermore, this introduction will challenge children to explore the ways in which their caring responsibility for the natural life–world is being regulated by social structures of a political, economic or cultural kind – including both constraints and opportunities. The identification of the social conditions of environmental responsibility may inspire children to act or speak in public discourse, aiming for a change of our collective practices of involvement with nature. As such, *education for environmental responsibility* aims to inspire children to develop a personal style of involvement with nature, simultaneously expressing their personal care in a way that brings about the promise of something new in public discourse on environmental issues.

As a curricular practice within elementary and secondary school education, it is suggested that environmental education consist of four types of activities. These activities are closely connected to each other, as represented by a matrix, positioning an axis of activities over and against an axis of loci:

Activity/locus	Local	Global
Expression	*Nature education*	*Exchange and participation*
Evaluation	*The process of taking responsibility*	*Global citizenship education*

Nature education serves as an umbrella term for educational activities taking place in a natural environment or with natural things (non-finished natural materials) that provide children with the opportunity to develop an intimate involvement with their natural life–world as expressed in a personal style of caring. Since nature education should enable children to follow their playful responses and use their free imagination, these activities are organised around open assignments and challenges, leaving maximum room for responsive exploration; rather than asking children to look out for a spider web and draw it, educators challenge them to look out for traces of insects, and show them to others.

Whereas the purpose of nature education is to sensitise children to the appeal originating from their caring involvement with nature, *the process of taking responsibility* is meant to stimulate children to reflect on their caring responses within everyday school practices and make them co-responsible for collective practices. Pupils are involved in the practical ways of dealing with garbage, the preparation of school lunch, the cleaning of the schoolyard, the organisation of school trips, the preservation of traffic rules for parents bringing and taking their children by car, as well as the collective evaluation of these practices within traditional school disciplines (ranging from history to economy and physics). In this sense, the process of taking responsibility is a matter of collective evaluation of shared practices by questioning each other's style of conduct and articulating the common codes of conduct sustaining these caring practices.

Then, there is the educational practice of *exchange and participation*. As children shift their attention from the school environment to abroad, the first step in 'carrying forward our acquaintanceship with the local into the global' can be made by expressing and sharing local experiences of environmental exploration together with children, scientists and politicians abroad. The possibilities offered by internet (e-mail groups, e-learning, e-meeting, e-publishing, e-government) provide them with the media to communicate with people from all over the world without immediate interference of official institutions and authorities. By exchanging local knowledge and experiences, children explore the global conditions of environmental responsibility. What are the institutional possibilities and constraints for environmental action? Thus, possibilities for collective action may open up and new forms of solidarity emerge around environmental issues on a global level (emission politics, fair international trade, energy policies, rules of global transport, development aid, water management, and agricultural support).

Finally, more or less disciplined reflection on previous experiences of international exchange and environmental action is likely to take place within *global citizenship education*. As such, this reflection will be sustained by curricular knowledge of environmental theory and international politics. Furthermore, environmental issues are discussed in close connection with issues of development, economic policies, social policies, international trade and current events in foreign affairs. Thus, joint reflection takes place within classroom discussion, structured around themes that pupils bring forward together with established curricular themes. Pupils are stimulated to examine the prevailing institutional practices for immanent but neglected and as (yet) unused possibilities to give new impulses to our caring responsibility for the natural life–world. However, the precise content, agenda and organisational form of these discussions cannot be settled in advance. Deliberation about these preliminary matters are the core responsibility of the pupils themselves, thus creating a democratic space in which they experience the pleasures and burdens of political action: a space for personal expression and judgement.

NOTES

[1] Dickens, C. (1854; 1995), pp. 11–12.

[2] www.globenederland.nl

[3] www.codenamefuture.nl

[4] In general terms, my proposal is probably similar to the proposal of Lucie Sauvé, labelled *Education for the Development of Responsible Societies* (as suggested in Sauvé, 1996, 1998 and 2002). At least we share a search for a different understanding of environmental responsibility: 'We should first distinguish between two conceptions of responsibility. There is the narrow one, associated with caution, respect, and the application of rules in a legalist framework; this is a shallow responsibility, which is instrumental and can be seen as having the characteristics of modernity with its individualist and anthropocentric focus. However, there is also a deeper responsibility or integral responsibility which shares some of the characteristics of reconstructive post modernity: a union of subject and object, of humans and nature (fundamental solidarity), between being and doing (authenticity), as well as consideration of the context of places and cultures where this responsibility is exercised. This second conception leads us to clarify the close connections between responsibility, consciousness, lucidity, reflectivity, freedom, autonomy, authenticity, commitment, courage, solidarity and care' (Sauvé,1998a). Furthermore, about environmental education she writes that: 'EE contributes to the development of responsible societies: an ethics of fundamental responsibility, that is significantly richer than the essentially minimalist ethics of sustainability ("so long as it lasts" or "so long as we survive"). The ethics of responsibility goes beyond a legalist and civic approach to rights and duties; it calls for a sense of responsibility for one's own being, knowledge and action, which implies commitment, lucidity, authenticity, solicitude and courage' (Sauvé, 2002).

BIBLIOGRAPHY

Achterberg, W. (1989). Identiteit en toekomstige generaties. *Algemeen Nederlands Tijdschrift voor Wijsbegeerte, 81*, 102–118.

Achterberg, W. (1989/1990). Duurzaamheid en intrinsieke waarde. *Wijsgerig Perspectief, 30*, 169–175.

Achterberg, W. (1994). *Samenleving, natuur en duurzaamheid*. Assen: Van Gorcum.

Achterberg, W. (1995). Can liberal democracy survive the environmental crisis? In: Dobson, A. and Lucardie, P. (eds) *The Politics of Nature* (pp. 113–146). London: Routledge.

Achterhuis, H. (1996). Weg met de utopieën. *NRC Handelsblad* (p. 14).

Altieri, C. (1994). *Subjective Agency. A Theory of First-Person Expressivity and Its Social Implications*. Oxford: Blackwell Publishers.

Apel, K. O. (1996). De ecologische crisis en het perspectief van de discours-ethiek. In: von Schomberg, R. (ed.) *Het discursieve tegengif. De sociale en ethische aspecten van de ecologische crisis* (pp. 31–53). Kampen: Kok Agora.

Arendt, H. (1958a). The crisis in education. *Partisan Review, 25*, 493–513.

Arendt, H. (1958b). *The Human Condition*. Chicago: The University of Chicago Press.

Arendt, H. (1961). *Between Past and Future*. New York: Viking Press.

Arendt, H. (1981). *The Life of the Mind*. New York: Harcourt Brace and Company.

Arendt, H. (1982). *Lectures on Kant's Political Philosophy*. Chicago: University of Chicago Press.

Bai, H. (1998). Autonomy reconsidered: a proposal to abandon the language of self- and other-control and to adopt the language of 'attunement'. In: *Philosophy of Education Society Yearbook 1998*. Illinois: University of Illinois.

Bakker, R. (1969). *Kerngedachten van merleau-ponty*. Roermond: Romen and zonen.

Barry, B. (1977). Justice between generations. In: Hacker, P. M. S. and Raz, J. (eds) *Law, Morality and Society* (pp. 267–284). Oxford: Oxford University Press.

Barry, B. (1978). Circumstances of justice and future generations. In: Sikora, R. L. and Barry, B. (eds) *Obligations to Future Generations* (pp. 204–248). Philadelphia: Temple University Press.

Bayles, M. D. (1980). *Morality and Population Policy*. Alabama: University of Alabama.

Beck, U. (1992). *Risk Society: Towards a New Modernity*. London: Sage.

Beck, U. (1998). *Democracy Without Enemies*. Cambridge: Polity Press.

Bell, D. R. (2004). Creating green citizens? Political liberalism and environmental education. *Journal of Philosophy of Education, 38*, 37–53.

Benedict, F. B. (ed.) (1991). *Environmental Education for Our Common Future*. Oslo: Norwegian University Press.

Beyer, A. (1998). *Nachhaltigkeit und Umweltbildung*. Hamburg: Krämer.

Biologieraad (2002). *Biologie: een vitaal belang*. The Hague: KNAW.

Blake, N., Smeyers, P., Smith, R. and Standish, P. (1998). *Thinking Again. Education After Postmodernism*. London: Bergin and Garvey.

Blake, N., Smeyers, P., Smith, R. and Standish, P. (2000). *Education in an Age of Nihilism*. London and New York: Routledge and Falmer.

Bolscho, D. (1998). Nachhaltigkeit – (k)ein Leitbild für Umweltbildung. In: Beyer, A. (ed.) *Nachhaltigkeit und Umweltbildung* (pp. 163–177). Hamburg: Krämer.

Bonnett, M. (1997). Environmental education and beyond. *Journal of Philosophy of Education, 31*, 249–266.

Bonnett, M. (2000). Environmental concern and the metaphysics of education. *Journal of Philosophy of Education, 34*, 591–602.

Bonnett, M. (2002). Education for sustainability as a frame of mind. *Environmental Education Research, 8*, 9–20.

Bonnett, M. (2003). Retrieving nature: education for a post-humanist age. *Journal of Philosophy of Education, 37 (special issue)*, 550–730.

Bonnett, M. and Cuypers, S. (2002). Autonomy and authenticity in education. In: Blake, N., Smeyers, P., Smith, R. and Standish, P. (eds) *The Blackwell Guide to Philosophy of Education* (pp. 326–340). Oxford: Blackwell Publishers.

Bowers, C. A. (2001). Challenges in educating for ecologically sustainable communities. *Educational Philosophy and Theory, 33*, 257–265.

Bowers, C. A. (2002). Towards an eco-justice pedagogy. *Environmental Education Research, 8*, 21–34.

Bransen, J. (1996). Identification and the idea of an alternative of oneself. *European Journal of Philosophy, 4*, 1–16.

Bransen, J. (2004). *Jezelf blijven. Inaugurale rede*. Nijmegen: KUN.

Burbules, N. and Smeyers, P. (2002). Wittgenstein, the practice of ethics and moral education. *Philosophy of Education Society*, Champaign, IL. (unpublished).

Burms, A. (2000). De plaats van het dier. *Tijdschrift voor Filosofie, 62*, 549–564.

Burms, A. and De Dijn, H. (1995). *De rationaliteit en haar grenzen. Kritiek en deconstructie*. Leuven: Universitaire Pers.

Callan, E. (1998). Autonomy and alienation. In: Hirst, P. H. and White, P. (eds) *Philosophy of Education. Major Themes in the Analytic Tradition. Vol. II. Education and Human Being* (pp. 68–93). New York and London: Routledge.

Carr, D. (2004). Moral values and the arts in environmental education: towards an ethics of aesthetic appreciation. *Journal of Philosophy of Education, 38*, 221–239.

Cheney, J. (1992). Intrinsic value in environmental ethics: beyond subjectivism and objectivism. *The Monist, 75*, 227–235.

Child, M., Williams, D. D. and Birch, A. J. (1995). Autonomy or heteronomy? Levinas's challenge to modernism and postmodernism. *Educational Theory, 45*, 167–189.

Coglianese, C. (1998). Implications of liberal neutrality for environmental policy. *Environmental Ethics, 20*, 41–59.

Cooper, D. E. (1998). Authenticity, life and liberal education. In: Hirst, P. H. and White, P. (eds) *Philosophy of Education. Major Themes in the Analytic Tradition. Vol. II. Education and Human Being* (pp. 32–67). New York and London: Routledge.

Cuypers, S.E. (1992). Is personal autonomy the first principle of education? *Journal of Philosophy of Education, 26*, 5–17.

Cuypers, S.E. (2000). Autonomy beyond voluntarism: in defense of hierachy. *Canadian Journal of Philosophy, 30*, 225–256.

Davidson, J. (2000). Sustainable development: business as usual or a new way of living? *Environmental Ethics, 22*, 25–42.

Dearden, R.F. (1998). 'Needs' in education. In: Hirst, P. H. and White, P. (eds) *Philosophy of Education. Major Themes in the Analytic Tradition. Vol. II. Education and Human Being* (pp. 255–267). New York and London: Routledge. (Origially published 1972).

DeLuca, K. M. (2001). Rethinking critical theory: instrumental reason, judgment, and the environmental crisis. *Environmental Ethics, 23*, 307–326.

De Haan, G. (2000). *Educating for Sutainability*. Berlin: Peter Lang.

De Jong, J. (1998). *Waardenopvoeding en Onderwijsvrijheid*. Nijmegen: Dissertatie.

De-Shalit, A. (1995). *Why posterity matters. Environmental policies and future generations*. London and New York: Routledge.

De Valck, M. (2004). Ieder kind heeft recht op zijn eigen bult. *63/6*, 11–13.

Den Boer, P. A. M. (1997). *Natuur- en Milieueducatie: Achtergronden, Kenmerken, Opvattingen*. Nijmegen: ITS.

Dickens, C. (1995). *Hard Times*. London: Penguin. (Original work published in 1854).
Dobson, A. (1998). *Justice and the Environment. Conceptions of Environmental Sustainability and Theories of Distributive Justice*. Oxford: Oxford University Press.
Drenthen, M. (1999). The paradox of environmental ethics; Nietzsche's view on nature and the wild. *Environmental Ethics, 21*, 163–175.
Drenthen, M. (2002). De wilde natuur en het verlangen naar andersheid, In: Voorsluis, B. (ed.) *Zwijgende natuur. Natuurervaring tussen betovering en onttovering* (pp. 65–86). Zoetermeer: Meinema.
Drenthen, M. (2003). *Grenzen aan Wildheid. Wildernisverlangen en de Betekenis van Nietzsches Moraalkritiek voor de Actuele Milieu-Ethiek*. Budel: Damon.
Dworkin, G. (1988a). Paternalism: some second thoughts. In: *The Theory and Practice of Autonomy* (pp. 150–160). Cambridge: Cambridge University Press.
Dworkin, G. (1988b). Behaviour control and design. In: *The Theory And Practice of Autonomy* (pp. 150–160). Cambridge: Cambridge University Press.
Dworkin, G. (1989). The concept of autonomy. In: Christman, J. (ed.) *The Inner Citadel*. New York and Oxford: Oxford University Press.
Edwards, R. and Usher, R. (2000). *Globalisation and Pedagogy*. London and New York: Routledge.
Elliot, R. (1992). Intrinsic value, enviromental obligation and naturalness. *The Monist, 75*, 138–160.
Emerson, W. (1883). *Nature, addresses and lectures*. Cambridge: Riverside Press.
Environmental Learning for the 21st Century. (1995). Paris: Centre for education research and innovation.
Feinberg, J. (1974a). The rights of animals and unborn generations. In: Blackstone, W.T. (ed.) *Philosophy and Environmental Crisis*. Athens: University of Georgia Press.
Feinberg, J. (1974b). Legal paternalism. *Canadian Journal of Philosophy, 1*, 105–124.
Fisher, J. A. (2001). Aesthetics. In: Jamieson, D. (ed.) *A Companion to Environmental Philosophy* (pp. 264–276). Malden and Oxford: Blackwell Publishers.
Frankfurt, H. G. (1988). *The Importance of What We Care About*. Cambridge: Cambridge University Press.
Frankfurt, H. G. (1999). *Necessity, Volition and Love*. Cambridge: Cambridge University Press.
Gadamer, H.-G. (1990). *Wahrheit und Methode. Grundzüge einer Philosophischen Hermeneutik*. Tübingen: J.C.B. Mohr. (Original work published in 1960).
Gauthier, D. (1963) *Practical Reasoning*. Oxford: Clarendon Press.
Gilbert, R. (1996). *Identity, Culture and Environment: Education for Citizenship for the 21st Century*. In: Demaine, J. and Entwistle, H. (eds) *Beyond Communitarianism: Citizenship, Politics, and Education* (pp. 42–63). Basingstoke: MacMillan Press.
Golding, M. P. (1972). Obligations to future generations. *The Monist, 56*, 85–99.
González-Gaudiano, E. (2001). Complexity in environmental education. *Educational Philosophy and Theory, 33*, 153–166.
Gordon, M. (1999). Hannah Arendt on authority: conservatism in education reconsidered. *Educational Theory, 49*, 161–180.
Gough, N. (1994). Playing the catastrophe: ecopolitical education after postmodernism. *Educational Theory, 44*, 189–209.
Gough, S. and Scott, W. (2001). Curriculum development and sustainable development: practices, institutions and literacies. *Educational Philosophy and Theory, 33*, 137–152.
Grey, W. (1996). Possible persons and the problems of posterity. *Environmental Values, 5*, 161–179
Gutman, A. and Thompson, D. (1990). Moral conflict and political consensus. In: Douglass, R. B., Mara, G. M. and Richardson, H. S. (eds) *Liberalism and The Good* (pp. 125–147). New York and London: Routledge.
Hargrove, E. C. (1989). *Foundations of Environmental Ethics*. New Jersey: Prentice Hall.
Hargrove, E. C. (1992). Weak anthropocentric intrinsic value. *The Monist, 75*, 183–207.
Held, V. (1988). Non-contractual society: a feminist view. In: Hanen, M. and Nielsen, K. *Science, Morality and Feminist Theory*. Calgary: University of Calgary Press.
Hertz, N. (2001). *The Silent Takeover. Global Capitalism and The Death of Democracy*. London: Arrow.
Hesselink, F., Kempen, P. P. and Wals, A. (2000). *ESDebate. International debate on Education for Sustainable Development*. Gland: IUCN Commission on Education and Communication.

Heyting, F. (1998). Opvoeden tot samenleven. Afscheid van moraal en deugd als voorwaarden voor maatschappelijke integratie. *Nederlands Tijdschrift voor Opvoeding, Vorming en Onderwijs, 14*, 35–49.

Heyting, F. (2004). Beware of ideals in education. *Journal of Philosophy of Education, 38*, 241–247.

Hilhorst, M. T. (1987). *Verantwoordelijk voor toekomstige generaties? Een sociaal-ethische bezinning op bevolkingsaantal, kernenergie, grondstoffen en genetica*. Kampen: Kok.

Hill, L. H. and Clover, D. E. (2003). *Environmental Adult Education. Ecological Learning, Theory, and Practice for Socio-Environmental Change*. San Francisco: Jossey-Bass.

Honohan, I. (2002). *Civic Republicanism*. London: Routledge.

Hood, R. (1998). Rorty and postmodern environmental ethics: recontextualizing narrative, reason and representation. *Environmental Ethics, 20*, 183–193.

Hopkins, C. and McKeown, R. (2001). Education for sustainable development: past experience, present action and future prospects. *Educational Philosophy and Theory, 33*, 231–244.

Howarth, R. B. (1992). Intergenerational justice and the chain of obligation. *Environmental Values, 1*, 133–140.

Huizinga, J. (1971). *Homo Ludens. A Study of the Play-Element in Culture*. Boston: The Beacon Press.

Humphrey, M. I. (1999). Deep ecology and the irrelevance of morality: a response. *Environmental Ethics, 21*, 75–79.

Husén, T. and Postlethwaite, T.N. (1994). *The International Encyclopedia of Education*. Oxford: Pergamon.

Hutchison, D. (1998). *Growing Up Green. Education for Ecological Renewal*. New York: Teachers College Press.

Imelman, J. D. (1974). *Plaats en inhoud van een personale pedagogiek*. University of Groningen. PhD-Thesis.

IUCN Commission in Education and Communication. (2004). *Supporting the United Nations decade on education for sustainable development 2005–2015*.

Gland: IUCN Commission in Education and Communication.

Jacobs, F. C. L. M. (1991). Kan de liberale democratie ons helpen de milieucrisis te overleven? In: Zweers, W. (ed.) *Op zoek naar een ecologische cultuur* (pp. 156–160). Baarn: Ambo.

Jans, M. and Wildemeersch (1999). Natuur- en milieu-educatie: van overtuigen naar overleggen. *Vorming, 14*, 161–180.

Jansen, T. (1994). *Gedeelde verschillen. Algemene volwassenenvorming in een veelvormige wereld*. Den Haag: VUGA.

Jickling, B. (1991). Environmental education and environmental advocacy: the need for a propper distinction. In: Lacey, C. and Williams, R. (eds) *To See Ourselves and To Save Ourselves: Ecology and Culture in Canada* (pp. 169–175). Montreal: Association for Canadian Studies.

Jickling, B. (1993). Research in environmental education: some thoughts on the need for conceptual analysis. *Australian Journal of Environmental Education, 9*, 85–94.

Jickling, B. (1994). Why I don't want my children to be educated for sustainable development. *The Journal of Environmental Education, 23*, 5–8.

Jickling, B. (1997). If environmental education is to make sense for teachers, we had better rethink how we define it! *Canadian Journal of Environmental Education, 2*, 86–103.

Jickling, B. (2001). Environmental thought, the language of sustainability, and digital wages. *Environmental Education Research, 7*, 167–180.

Jickling, B. and Spork, H. (1998). Education for the environment: a critique. *Environmental Education Research, 4*, 309–327.

Jonas, H. (1984). *The Imperative of Responsibility. In: Search of an Ethics for the Technological Age*. Chicago and London: The University of Chicago Press.

Kant, I. (1987). *The Critique of the Power of Judgement* (W. Pluhar, Trans.). Cambridge: Hackett. (Original work publsihed 1790).

Kavka, G. S. (1981). The paradox of future individuals. *Philosophy and Public Affairs, 11*, 93–112.

Klein, N. (2000). *No Logo*. London: Flamingo.

Klop, C. J. (1993). *De cultuurpolitieke paradox. Noodzaak èn onwenselijkheid van overheidsinvloed op normen en waarden*. Kampen: Kok.

Koelega, D. G. A. (1995). Technology, ecology, autonomy and the state. *Techné, 1*, 7–13.

Koelega, D. G. A. (2000). *Niet alles van waarde is vervangbaar. Een liberaal perspectief op behoud van natuur en cultuurgoederen in een technologische samenleving*. Kampen: Kok.

Korthals, M. (1994). *Duurzaamheid en democratie. Sociaal-filosofische beschouwingen over milieubeleid, wetenschap en technologie*. Amsterdam and Meppel: Boom.

Kymlicka, W. and Wayne, N. (1994). Return of the citizen: a survey of recent work on citizenship theory. *Ethics, 104*, 352–381.

Landelijk Ambitiestatement Programma Leren voor Duurzaamheid (2000). Den Haag: Stuurgroep Leren voor Duurzaamheid.

Langeveld, M. J. (1963). *Beknopte theoretische paedagogiek*. Groningen: Wolters.

Lemaire, T. (2002). *Met open zinnen. Natuur, landschap, aarde*. Amsterdam: Ambo.

Leren voor Duurzame Ontwikkeling 2004–2007. Van Marge naar Mainstream (2003). Den Haag: Stuurgroep Leren voor Duurzaamheid.

Les Brown, M. (1987). *Conservation and Practical Morality. Challenges to Education and Reform*. New York: St Martins Press.

Levinson, N. (1997). Teaching in the midst of belatedness: the paradox of natality in Hannah Arendt's educational thought. *Educational Theory, 47*, 435–451.

Li, H. (1994). Environmental education: rethinking intergenerational relationship. In: *Philosophy of Education Society Yearbook 1996*. Illinois: University of Illinois.

Li, H. (1996). On the nature of environmental education. Anthropocentrism versus non-anthropocentrism: the irrelevent debate. In: *Philosophy of Education Society Yearbook 1996*. Illinois: University of Illinois.

Lijmbach, S., Broens, M. and Hovinga, D. (2000). *Duurzaamheid als leergebied*. Utrecht: CDß-Press.

Lijmbach, S., Margadant-Van Arcken, M., van Koppen, C. S. A. and Wals, A. E. J. (2002). 'Your view of nature is not mine!': learning about pluralism in the classroom. *Environmental Education Research, 8*, 121–135.

Luke, T. W. (2001). Education, environment and sustainability: what are the issues, where to intervene, what must be done? *Educational Philosophy and Theory, 33*, 187–202.

Lyotard, J. F. (1984). *The Postmodern Condition: A Report on Knowledge*. Manchester: Manchester University Press.

MacIntyre, A. and Dunne, J. (2002). Alasdair MacIntyre on education: In dialogue with Joseph Dunne. *Journal of Philosophy of Education, 36*, 1–19.

Magill, K. (2000). Blaming, understanding and justification. In: van den Beld, T. (ed.) *Moral Responsibility and Ontology*. Dordrecht: Kluwer.

Manenschijn, G. (1992). The preservation of nature and the dominance of anthropocentrism in liberal ethics. In: Musschenga, A.W., Voorzanger, B. and Soeteman, A. (eds) *Morality, Worldview and Law* (pp. 255–266). Assen: Van Gorcum.

Manning, R. (1981). Environmental ethics and John Rawls' theory of justice. *Environmental Ethics, 3*, 155–165.

Margadant-van Arcken, M. (1990). *Groen verschiet. Natuurbeleving en natuuronderwijs bij acht- tot twaalfjarige kinderen*. Den Haag: SDU Uitgeverij.

Margadant-van Arcken, M. (1998). *Rehabilitatie van Leefwerelddenken*. Wageningen: Wageningen University Press.

Marsden, W. E. (1997). Environmental education: historical roots, comparative perspectives, and current issues in Britain and the United States. *Journal of Curriculum and Supervision, 13*, 6–29.

Marshall, J. D. (1995). Foucault and neo-liberalism: biopower and busno-power. In: *Philosophy of Education Society Yearbook 1995*. Illinois: University of Illinois.

Matthews, F. (2001). Deep ecology. In: Jamieson, D. (ed.) *A Companion to Environmental Philosophy* (pp. 218–231). Malden (Mass.) and Oxford: Blackwell Publishers.

McKeown, R. (2002). *Education for Sustainable Development Toolkit*. Knoxville: University of Tennessee.

McKeown, R. and Hopkins, C. (2003). EE ≠ ESD: defusing the worry. *Environmental Education Research, 9*, 117–128.

McKibben, W. (1989). *The End of Nature*. New York: Random House.

Meijer, W. A. J. (1996). Milieu-educatie en het milieuprobleem: een cultuurpedagogisch exempel. *Stromingen in de pedagogiek* (pp. 125–145). Baarn: Intro.

Meijer, W. A. J. (1997). Maatschappelijke problemen horen niet in de school. Waardenoriënterend onderwijs en de dubbele moraal. *Vernieuwing, 56/2*, 10–13.

Merleau-Ponty, M. (1962). *Phenomenology of Perception* (C. Smith, Trans.). London: Routledge and Kegan Paul. (Original work published 1945).

Merleau-Ponty, M. (1997). *The visible and the Invisible* (A. Lingis, Trans.). Evanston: Northwestern University Press. (Original work published 1964).

Mill, J. S. (1859). *On Liberty*. London: Packer.

Ministerie van Verkeer, Ruimtelijke Ordening and Milieu. (1987). *Nota natuur- en milieu-educatie: een meerjarenplan*. Den Haag: SDU Uitgeverij.

Mollenhauer, K. (1986). *Vergeten samenhang. Over opvoeding en cultuur*. Amsterdam: Boom.

Mouffe, C. (2000). *The Democratic Paradox*. London and New York: Verso.

Musschenga, A. W. (1991). Liberale neutraliteit en de rechtvaardiging van milieubehoud. In: Zweers, W. (ed.) *Op zoek naar een ecologische cultuur* (pp. 161–170). Baarn: Ambo.

Naess, A. (1989). *Ecology, Community and Lifestyle*. Cambridge: Cambridge University Press.

Naess, A. and Jickling, B. (2000). Deep ecology and education: a conversation with Arne Naess. *Environmental Education Research, 5*, 48–62.

Nietzsche, F. (1982). *Daybreak. Thoughts on the Prejudices of Morality* (R. J. Hollingdale, Trans.). Cambridge: Cambridge University Press. (Original work published 1881).

Noddings, N. (1984). *Caring. A Feminine Approach to Ethics and Moral Education*. Berkeley, Los Angeles and London: University of California Press.

Noddings, N. (1998). Caring. In: Hirst, P. H. and White, P. (eds) *Philosophy of Education. Major Themes in the Analytic Tradition. Vol. II. Education and Human Being* (pp. 40–50). New York and London: Routledge.

Noddings, N. and Slote, M. (2002). Changing notions of the moral and of moral education. In: Blake, N., Smeyers, P., Smith, R. and Standish, P. (eds) *The Blackwell Guide to Philosophy of Education* (pp. 341–355). Oxford: Blackwell Publishers.

Norton, B. G. (1992). Epistemology and environmental values. *The Monist, 75*, 208–226.

Notitie NME 21 (1999). Tweede Kamer. Den Haag: SDU Uitgeverij.

Okin, S.M. (1990). Feminism, the individual, and contract theory. *Ethics, 100*, 658–669.

Oldfield, A. (1990). *Citizenship and Community. Civic Republicanism and the Modern World*. London and New York: Routledge.

O'Loughlin, M. (1995). Intelligent bodies and ecological subjectivities: Merleau-Ponty's corrective to postmodernism's 'subjects' of education. In: *PES Yearbook 1995*. Illinois: University of Illinois.

O'Loughlin, M. (2002). Embodied place as a source of national identity: a challenge to *ethnie* as the basis for citizenship and a potential for civic education. Paper presented at the *Conference Proceedings of the International Network of Philosophers of Education 8th Biennial Conference on The Many Faces of Philosophy of Education: Traditions, Problems and Challenges*. Oslo: University of Oslo. (Unpublished).

O' Neill, J. (1992). The varieties of intrinsic value. *The Monist, 75*, 119–137.

O'Neill, J. (2001). Meta-ethics. In: Jamieson, D. (ed.) *A Companion to Environmental Philosophy* (pp. 163–176). Malden and Oxford: Blackwell Publishers.

Otten, W. J. (1998). *Eindaugustuswind*. Amsterdam: Van Oorschot.

Parfit, D. (1984). *Reasons and Persons*. Oxford: Clarendon Press.

Passmore, J. (1974). *Man's Responsibility for Nature*. London: Duckworth.

Peters, M. (2001). Environmental education, neo-liberalism and globalisation: the 'New Zealand experiment'. *Educational Philosophy and Theory, 33*, 203–216.

Peters, R. S. (1998). Freedom and the development of the free man. In: Hirst, P. H. and White, P. (eds) *Philosophy of Education. Major Themes in the Analytic Tradition. Vol. II. Education and Human Being* (pp. 68–93). New York and London: Routledge. (Originally published 1973).

Pettit, P. (1997). *Republicanism. A Theory of Freedom and Government*. Oxford: Oxford University Press.

Postma, D. W. (2001). Over de noodzaak én onwenselijkheid van een opvoeding tot 'milieuvriendelijk' burgerschap. In: Smeyers, P. and Levering, B. (eds) *Grondslagen van de wetenschappelijke pedagogiek. Modern en postmodern* (pp. 208–227). Amsterdam: Boom.

Postma, D. W. (2002). Taking the future seriously. On the inadequacies of the framework of liberalism for environmental education. *Journal of Philosophy of Education, 36*, 41–56.

Postma, D. W. (2003). Personal autonomy, authenticity and the intrinsic valuation of nature. In: Smeyers, P. and Depaepe, M. (eds) *Beyond Empiricism. On Criteria for Educational Research* (pp. 237–246). Leuven: Leuven University Press.

Postma, D. W. and Emans, B. (2004). Milieu-educatie op zoek naar inspiratie. *Vernieuwing, 63/3*, 4–7.

Praamsma, J. M. (1993). Natuur en milieu-educatie, tussen beleven en overleven. *Pedagogisch Tijdschrift, 18*, 139–145.

Praamsma, J. M. (1997). *Nieuwe wereldburgers. Aantasting van natuur en milieu als vraagstuk van algemene vorming. Een zaakpedagogiek*. Utrecht: Dissertation.

Ramaekers, S. (2001). Feministisch onderzoek in de pedagogiek. In: Smeyers, P. and Levering, B. (eds) *Grondslagen van de Wetenschappelijke Pedagogiek* (pp. 307–326). Amsterdam: Boom.

Ramaekers, S. and Smeyers, P. (2006). Child rearing: passivity and being able to go on. Wittgenstein on shared practices and seeing aspects. *Journal of Philosophy of Education*.(in press)

Rawls, J. (1971). *A Theory of Justice*. Cambridge: The Belknap Press of Harvard University Press.

Rawls, J. (1993). *Political Liberalism*. New York: Columbia University Press.

Rawls, J. (1999). *A Theory of Justice. Revised edition*. Cambridge: The Belknap Press of Harvard University Press.

Regan, T. (1992). Does environmental ethics rest on a mistake? *The Monist 75*, 161–182.

Reid, A. (2002a). Discussing the possibility of education for sustainable development. *Environmental Education Research, 8*, 73–79.

Reid, A. (2002b). On the possibility of education for sustainable development. *Environmental Education Research, 8*, 5–7.

Rolston III, H. (1982). Are values in nature subjective are objective? *Environmental Ethics, 4*, 125–151.

Rolston III, H. (1992). Disvalues in nature. *The Monist, 75*, 250–278.

Rorty, R. (1980). *Philosophy and the Mirror of Nature*. Oxford: Blackwell Publishers.

Sagoff, M. (1988). Can environmentalists be liberals? In: Sagoff, M. (ed.) *The Economy of the Earth. Philososphy, Law and the Environment* (pp. 146–170). Cambridge: Cambridge University Press.

Sandel, M. (1982). *Liberalism and the Limits of Justice*. Cambridge: Cambridge University Press.

Sankowski, E. (1998). Autonomy, education, and politics. In: *Philosophy of Education Society Yearbook 1998*. Illinois: University of Illinois.

Sauvé, L. (1996). Environmental education and sustainable development: a further Appraisal. *Canadian Journal of Environmental Education, 1*, 7–33.

Sauvé, L. (1998). Environmental education between modernity and postmodernity: searching for an integrating educational framework. In: Jarnet, A., Jickling, B., SauvÈ, L., Wals, A. and Clarkin, P. (eds), *The Future of Environmental Education in a Postmodern World?* (pp. 57–70). Yukon: University Press.

Sauvé, L. (2002). Environmental education: possibilities and constraints. *Connect, 17*, 1–4.

Schalow, F. (2002). Who speaks for the animals? Heidegger and the question of animal welfare. *Environmental Ethics, 22*, 259–271.

Schwartz, T. (1978). Obligations to posterity. In: Sikora, R. L. and Barry, B. (eds) *Obligations to Future Generations*. Philadelphia: Temple University.

Scott, W. and Oulton, C. (1998). Environmental values education: an exploration of its role in the school curriculum. *Journal of Moral Education, 27*, 209–224.

Scruton, R. (1998). *Animal Rights and Wrongs*. London: Demos.

Singer, B. A. (1988). An extension of Rawls' theory of justice to environmental ethics. *Environmental Ethics, 10*, 217–231.

Singer, P. (1993). *Practical Ethics*. Cambridge: Cambridge University Press.

Signaaladvies natuur- en milieu-educatie. (1992). Den Haag: Raad voor het Milieu- en Natuuronderzoek.

Smeyers, P. (1997). Besprekingsartikel 'subjective agency' van Charles Altieri. *Pedagogisch Tijdschrift, 22*, 149–166.

Smeyers, P. (2001). Work in philosophy and work on oneself. Wittgenstein and Nietzsche: a balanced educational inheritance. Paper presented at *the Meeting of the Philosophy of Education Society of the USA*, Chicago, USA.

Smeyers, P. and Masschelein, J. (2000a). Contouren en uitwegen van de representatiecrisis in de pedagogiek. *Pedagogische Tijdschrift, 25*, 213–232.

Smeyers, P. and Masschelein, J. (2000b). L'enfance, education, and the politics of meaning. In: Dhillon, P. A. and Standish, P. (eds), *Lyotard. Just Education* (pp. 140–156). London and New York: Routledge.

Smith, M. J. (1998a). *Ecologism. Towards Ecological Citizenship*. Buckingham: Open University Press.

Smith, P. K and Vollstedt, R. (1985). Defining play: an empirical study of the relationship between play and various play criteria. *Child Development, 56*, 1042–1050.

Smith, R. (1998b). The education of autonomous citizens. In: Bridges, D. (ed.) *Education, Autonomy and Democratic Citizenship* (pp. 127–137). London: Routledge.

Smith, R. (1998c). Spirit of middle earth: practical thinking for an instrumental age. In: Cooper, D. and Palmer, J. (eds) *Spirit of the Environment* (pp. 168–181). London: Routledge.

Snauwaert, D. T. (1996). The relevance of the anthropocentric-ecocentric debate. In: *Philosophy of Education Society Yearbook 1996*. Illinois: University of Illinois.

Snik, G. (1991). Milieucrisis, antropocentrismekritiek, opvoeding. In: Heyting, F. et al. *Individuatie en socialisatie in tijden van modernisering* (pp. 90–97). Amsterdam: Siswo.

Snik, G. (1999). Grondslagen van liberale visies op staatsbemoeiing met onderwijs. *Pedagogisch Tijdschrift, 24*, 125–151.

Snik, G. and van Haaften, W. (1992). Moral autonomy as an aim of education. In: Musschenga, A.W., Voorzanger, B. and Soeteman, A. (eds) *Morality, Worldview and Law* (pp. 137–147). Assen: Van Gorcum.

Stables, A. (2001). Who drew the sky? Conflicting assumptions in environmental education. *Educational Philosophy and Theory, 33*, 245–256.

Stables, A. and Scott, W. (2001). Editorial. *Educational Philosophy and Theory, 33*, 133–135.

Standish, P. (2003). Levinas' language and the language of the curriculum. In Smeyers, P. and Depaepe, M. (eds), *Philosophy and History of the Discipline of Education. Evaluation and Evolution of the Criteria for Educational Research* (pp. 107–112). Leuven: University of Leuven.

Strawson, P. F. (1974). *Freedom and Resentment*. London: Methuen and Co. Ltd.

Taylor, C. (1969). Responsibility for self. In: Oksenberg, R. A. (ed.) *The Identities of Persons*. Berkeley: University of California Press.

Taylor, C. (1989). *Sources of the Self. The Making of the Modern Identity*. Cambridge: Cambridge University Press.

Taylor, C. (1992). *The Ethics of Authenticity*. Cambridge, Mass.: Harvard University Press.

Taylor, C. (1995). *Philosophical Arguments*. Cambridge (Mass.):Harvard University Press.

The United Nations Environment Programme (1972). *Declaration of the United Nations Conference on the Human Environment*. Stockholm: UNCED, UN Publications.

The United Nations Economic Commission for Europe. (2003). *Draft Statement on Education for Sustainable Development by the UNECE Ministers of the Environment, submitted by the Committee on Environmental Policy of the UNECE, Kiev, Ukraine*.

Thero, D. P. (1995). Rawls and environmental ethics: a critical examination of the literature. *Environmental Ethics, 17*, 93–106.

Thiel, F. (1996). *Ökologie als Thema. Überlegungen zur Pädagogisierung einer Gesellschaftlichen Krisenerfahrung*. Weinheim: Deutscher Studien Verlag.

Van der Wal, G. A. (1979). *De Zorg Voor Lateren: Van Latere Zorg?* Rotterdam: Erasmus Universiteit.

United Nations Conference on Environment and Development. (1992). *Agenda 21: Report of the United Nations Division for Sustainable Development*. 3–14 June 1992, Rio de Janeiro, Brazil. New York: Oceana Publications.

United Nations Conference on Environment And Development. (1992). *Report of the commission of the European communities to the United Nations conference on environment and development*. Rio de Janeiro: UNCED.

United Nations Educational Social and Cultural Organisation. (1992) *Reshaping Education for Sustainable Development*. Paris: UNESCO.

United Nations Commission on Sustainable Development. (2001). *Report of the Secretary-General on Education And Public Awareness for Sustainable Development* (Report no. E/CN.17/2001/PC/7), New York, NY: UNESCO, UN Publications.

Van Asperen, G. M. (1993). *Het bedachte leven*. Amsterdam and Meppel: Boom.

Vanderstraeten, R. and Biesta, G. J. J. (2001). How is education possible? Preliminary investigations for a theory of education. *Educational Philosophy and Theory, 33*, 7–21.

Van Haaften, A.W. (1979). *Epistemologisch relativisme. Logisch en psychologisch perspectief in de filosofische argumentatie*. Leiden: PhD-thesis.

Van Haaften, A.W. (1996). Relativism and absolutism: how both can be right. *Metaphilosophy, 27*, 324–326.

Van der Wal, G. A. (1989/1990). De zorg om de lateren, een oud thema in nieuw gewaad. *Wijsgerig Perspectief, 30*, 158–163.

Van der Wal, G. A. (2002). Hans Jonas' filosofie van het organische. *Filosofie, 12*, 4–10.

Van Gunsteren, H. R. (1992). *Eigentijds Burgerschap. WRR-Publikatie*. Den Haag: Sdu Uitgeverij.

Van Putten, J. (1999). Stel je voor dat je in de eindtijd leeft. *Filosofie Magazine, 8*, 42–45.

Vansieleghem, N. (2004). Niets is voor altijd. Opvoeding als verbeeldende herinnering. Een aanzet tot herdenking van het voorbeeld in opvoeding. In: Leyder, D., Vanobbergen, B. and Vansieleghem, N. (eds) *Opvoeding. Een (on)persoonlijke aangelegenheid? Wijsgerige en historische reflecties op de plaats van de opvoeder in het educatief gebeuren* (pp. 71–77). Amsterdam: SWP.

Van Zomeren (1998). *De bewoonde wereld. Een bloemlezing uit ons landschap*. Amsterdam: Arbeiderspers.

Vermeersch, E. (1995). Educatieve implicaties van een globale visie op de milieuproblematiek. In: van Bergeijk, J., Alblas, A.H. and Visser-Reyneveld, M.I. (eds) *Natuur- en milieu-educatie didactisch beschouwd* (pp. 145–153). Wageningen: Wageningen Pers.

Vincent, A. (1998). Liberalism and the environment. *Environmental Values, 7*, 443–459.

Visser 't Hooft, H. Ph. (1996). De ecologische crisis en de plicht tot overleven. In: von Schomberg, R. (ed.) *Het discursieve tegengif. De sociale en ethische aspecten van de ecologische crisis* (pp. 54–73). Kampen: Kok Agora.

Visser 't Hooft, H. Ph. (2002). Hans Jonas en onze verantwoordelijkheid voor de toekomst. *Filosofie, 12*, 11–16.

Vogel, S. (2002). Environmental philosophy after the end of nature. *Environmental Ethics, 24*, 23–39.

Vokey, D. (1994). Education for intergenerational justice: why should we care? In *Philosophy of Education Society Yearbook 1996*. Illinois: University of Illinois.

Waks, L. J. (1996). Environmental claims and citizen rights. *Environmental Ethics, 18*, 133–148.

WCED (1987). Our common future: the World Commission On Environment and Development. Oxford: Oxford University Press

Wenz, P. S. (1988). *Environmental justice*. New York: SUNY.

Weston, A. (1992). Between means and ends. *The Monist, 75*, 236–249.

Whiteside, K. H. (1994). Hannah Arendt and ecological politics. *Environmental Ethics, 16*, 339–358.

Whiteside, K. H. (1998). Worldliness and respect for nature: an ecological application of Hannah Arendt's conception of culture. *Environmental values, 7*, 25–40.

Wijffels, B. (2004). Ondersteuning. Trends in natuur- en milieueducatie. *Vernieuwing, 63/3*, 17–20.

Winch, P. (1958). *The Idea of a Social Science*. London: Routledge and Kegan Paul.

Wittgenstein, L. (1923). *Tractatus logico-philosophicus*. London: Kegan Paul.

Wittgenstein, L. (1958). *Philosophical investigations*. Oxford: Blackwell Publishers.

Zwart, H. (1997). What is an animal? A philosophical reflection on the possibility of a moral relationship with animals. *Environmental Values, 6*, 377–392.

INDEX

The International Library of Environmental, Agricultural and Food Ethics

1. H. Maat: *Science Cultivating Practice.* A History of Agricultural Science in the Netherlands and its Colonies, 1863-1986. 2002 ISBN 1-4020-0113-4
2. M.K. Deblonde: *Economics as a Political Muse.* Philosophical Reflections on the Relevance of Economics for Ecological Policy. 2002 ISBN 1-4020-0165-7
3. J. Keulartz, M. Korthals, M. Schermer, T.E. Swierstra (eds.): *Pragmatist Ethics for a Technological Culture.* 2003 ISBN 1-4020-0987-9
4. N.P. Guehlstorf: *The Political Theories of Risk Analysis.* 2004 ISBN 1-4020-2881-4
5. M. Korthals: *Before Dinner.* Philosophy and Ethics of Food. 2004 ISBN 1-4020-2992-6
6. C.J. Preston and W. Ouderkirk (eds.): *Nature, Value, Duty.* Life on Earth with Holmes Rolston, III. 2007 ISBN 1-4020-4877-7
7. D.W. Postma: *Why Care for Nature?* In Search of an Ethical Framework for Environmental Responsibility and Education. 2006 ISBN 1-4020-5002-X

DATE DUE

Demco, Inc. 38-293